# Advanced Courses in Mathematics
# CRM Barcelona

Centre de Recerca Matemàtica

*Managing Editor:*
Carles Casacuberta

Ieke Moerdijk • Bertrand Toën

# Simplicial Methods for Operads and Algebraic Geometry

*Editors for this volume:*
Carles Casacuberta (Universitat de Barcelona)
Joachim Kock (Universitat Autònoma de Barcelona)

 Birkhäuser

Ieke Moerdijk
Mathematisch Instituut
Universiteit Utrecht
Postbus 80.010
3508 TA Utrecht
The Netherlands
e-mail: I.Moerdijk@uu.nl

Bertrand Toën
I3M UMR 5149
Université Montpellier 2
Case Courrier 051
Place Eugène Bataillon
34095 Montpellier Cedex
France
e-mail: btoen@math.univ-montp2.fr

2010 Mathematics Subject Classification: primary: 55U40, 18G30;
secondary: 55P48, 18D50, 18F10, 14A20

ISBN 978-3-0348-0051-8       e-ISBN 978-3-0348-0052-5
DOI 10.1007/978-3-0348-0052-5

*Cover design*: deblik, Berlin

Printed on acid-free paper

Springer Basel AG is part of Springer Science+Business Media

www.birkhauser-science.com

# Foreword

This book is an introduction to two higher-categorical topics in algebraic topology and algebraic geometry relying on simplicial methods. It is based on lectures delivered at the Centre de Recerca Matemàtica in February 2008, as part of a special year on Homotopy Theory and Higher Categories.

Ieke Moerdijk's lectures constitute an introduction to the theory of *dendroidal sets*, an extension of the theory of simplicial sets designed as a foundation for the homotopy theory of operads. The theory has many features analogous to the theory of simplicial sets, but it also reveals many new phenomena, thanks to the presence of automorphisms of trees. Dendroidal sets admit a closed symmetric monoidal structure related to the Boardman–Vogt tensor product. The lecture notes develop the theory very carefully, starting from scratch with the combinatorics of trees, and culminating with a model structure on the category of dendroidal sets for which the fibrant objects are the inner Kan dendroidal sets. The important concepts are illustrated with detailed examples.

The lecture series by Bertrand Toën is a concise introduction to *derived algebraic geometry*. While classical algebraic geometry studies functors from the category of commutative rings to the category of sets, derived algebraic geometry is concerned with functors from simplicial commutative rings (to allow derived tensor products instead of the usual ones) to simplicial sets (to allow derived quotients instead of the usual ones). The central objects are derived (higher) stacks, which are functors satisfying a certain up-to-homotopy descent condition. The lectures start with motivating examples from moduli theory, to move on to simplicial presheaves and algebraic (higher) stacks; next comes the homotopy theory of simplicial commutative rings, and finally everything comes together in the notion of derived (higher) stack. Some proofs are given as exercises that involve consulting the literature.

Both lecture series assume a working knowledge of Quillen model categories. For Toën's lectures, some background in algebraic geometry à la Grothendieck is also necessary.

We are very thankful to the CRM for hosting the advanced course as well as the whole research programme on Homotopy Theory and Higher Categories. The former director Manuel Castellet and his successor Joaquim Bruna made this possible. The CRM secretaries were much more than helpful at all times. We are also very much indebted to the programme co-organisers, André Joyal, Amnon Neeman, and Frank Neumann.

We acknowledge financial support from the i-MATH programme (Ingenio Mathematica, Consolider – Ingenio 2010) under grant PMII-C2-0055, the Catalan Government (Generalitat de Catalunya) under grant 2007/ARCS00104, and the Spanish Ministry of Science and Innovation under grant MTM2007-29363-E.

Above all, we thank, of course, the two authors, for their expertise, patience and kind collaboration.

Carles Casacuberta and Joachim Kock

# Contents

Foreword      v

**I   Lectures on Dendroidal Sets**
*Ieke Moerdijk*      **1**

Preface      **3**

**1   Operads**      **5**
   1.1   Operads . . . . . . . . . . . . . . . . . . . . . . . . . . . . . . .   5
   1.2   Coloured operads . . . . . . . . . . . . . . . . . . . . . . . . . .   7
   1.3   Examples of coloured operads . . . . . . . . . . . . . . . . . . .   8

**2   Trees as operads**      **11**
   2.1   A formalism of trees . . . . . . . . . . . . . . . . . . . . . . .   11
   2.2   Planar trees . . . . . . . . . . . . . . . . . . . . . . . . . . . . .   12
      2.2.1   Face maps . . . . . . . . . . . . . . . . . . . . . . . . . .   14
      2.2.2   Degeneracy maps . . . . . . . . . . . . . . . . . . . . . . .   14
      2.2.3   Dendroidal identities . . . . . . . . . . . . . . . . . . . . .   15
   2.3   Non-planar trees . . . . . . . . . . . . . . . . . . . . . . . . . .   17
      2.3.1   Dendroidal identities with isomorphisms . . . . . . . . . .   20
      2.3.2   Isomorphisms along faces and degeneracies . . . . . . . .   20
      2.3.3   The presheaf of planar structures . . . . . . . . . . . . . .   21
      2.3.4   Relation with the simplicial category . . . . . . . . . . . .   21

**3   Dendroidal sets**      **23**
   3.1   Basic definitions and examples . . . . . . . . . . . . . . . . . .   23
   3.2   Faces, boundaries and horns . . . . . . . . . . . . . . . . . . . .   29
   3.3   Skeleta and coskeleta . . . . . . . . . . . . . . . . . . . . . . .   32
   3.4   Normal monomorphisms . . . . . . . . . . . . . . . . . . . . . .   35

**4  Tensor product of dendroidal sets**                                              **41**
   4.1  The Boardman–Vogt tensor product . . . . . . . . . . . . . . .   41
   4.2  Tensor product of dendroidal sets . . . . . . . . . . . . . . .   44
   4.3  Shuffles of trees . . . . . . . . . . . . . . . . . . . . . . . .   47

**5  A Reedy model structure on dendroidal spaces**                                   **55**
   5.1  Strict Reedy categories . . . . . . . . . . . . . . . . . . . .   55
   5.2  Model structures for strict Reedy categories . . . . . . . . . .   56
   5.3  Generalized Reedy categories . . . . . . . . . . . . . . . . .   57
   5.4  Model structures for generalized Reedy categories . . . . . . .   60
   5.5  Dendroidal objects and simplicial objects . . . . . . . . . . .   63
   5.6  Dendroidal Segal objects . . . . . . . . . . . . . . . . . . . .   65

**6  Boardman–Vogt resolution and homotopy coherent nerve**                           **69**
   6.1  The classical $W$-construction . . . . . . . . . . . . . . . . .   69
   6.2  The generalized $W$-construction . . . . . . . . . . . . . . . .   74
   6.3  The homotopy coherent nerve . . . . . . . . . . . . . . . . . .   75

**7  Inner Kan complexes and normal dendroidal sets**                                 **79**
   7.1  Inner Kan complexes . . . . . . . . . . . . . . . . . . . . . .   79
   7.2  Inner anodyne extensions . . . . . . . . . . . . . . . . . . .   82
   7.3  Homotopy in an inner Kan complex . . . . . . . . . . . . . . .   84
   7.4  Homotopy coherent nerves are inner Kan . . . . . . . . . . . .   88
   7.5  The exponential property . . . . . . . . . . . . . . . . . . .   91

**8  Model structures on dendroidal sets**                                            **93**
   8.1  Preliminaries . . . . . . . . . . . . . . . . . . . . . . . . .   93
      8.1.1  Tensor product . . . . . . . . . . . . . . . . . . . .   96
      8.1.2  Intervals . . . . . . . . . . . . . . . . . . . . . . .   97
      8.1.3  Normalization . . . . . . . . . . . . . . . . . . . . .   97
   8.2  A Quillen model structure on planar dendroidal sets . . . . . .   99
   8.3  Trivial cofibrations . . . . . . . . . . . . . . . . . . . . . .  102
   8.4  A Quillen model structure on dendroidal sets . . . . . . . . . .  107

**Bibliography**                                                                     **117**


**II  Simplicial Presheaves and Derived Algebraic Geometry**
   *Bertrand Toën*                                                        **119**

**1  Motivation and objectives**                                                     **121**
   1.1  The notion of moduli spaces . . . . . . . . . . . . . . . . . .  121
   1.2  Construction of moduli spaces: one example . . . . . . . . . .  122
   1.3  Conclusions . . . . . . . . . . . . . . . . . . . . . . . . . .  126

**2 Simplicial presheaves as stacks** **127**
  2.1 Review of the model category of simplicial presheaves . . . . . . . 127
  2.2 Basic examples . . . . . . . . . . . . . . . . . . . . . . . . . . . . 133

**3 Algebraic stacks** **143**
  3.1 Schemes and algebraic $n$-stacks . . . . . . . . . . . . . . . . . . 144
  3.2 Some examples . . . . . . . . . . . . . . . . . . . . . . . . . . . . 147
  3.3 Coarse moduli spaces and homotopy sheaves . . . . . . . . . . . . 154

**4 Simplicial commutative algebras** **159**
  4.1 Quick review of the model category of commutative simplicial
      algebras and modules . . . . . . . . . . . . . . . . . . . . . . . 159
  4.2 Cotangent complexes . . . . . . . . . . . . . . . . . . . . . . . . 161
  4.3 Flat, smooth and étale morphisms . . . . . . . . . . . . . . . . . 163

**5 Derived stacks and derived algebraic stacks** **167**
  5.1 Derived stacks . . . . . . . . . . . . . . . . . . . . . . . . . . . . 167
  5.2 Algebraic derived $n$-stacks . . . . . . . . . . . . . . . . . . . . . 171
  5.3 Cotangent complexes . . . . . . . . . . . . . . . . . . . . . . . . . 175

**6 Examples of derived algebraic stacks** **179**
  6.1 The derived moduli space of local systems . . . . . . . . . . . . . 179
  6.2 The derived moduli of maps . . . . . . . . . . . . . . . . . . . . . 183

**Bibliography** **185**

# Part I

# Lectures on Dendroidal Sets

## Ieke Moerdijk

Notes written by Javier J. Gutiérrez

# Preface

The theory of dendroidal sets forms a new attempt to give a combinatorial theory of topological structures which involve operations with one output and multiple inputs, such as the theory of operads. The theory seems to be developing in a way similar to that of simplicial sets, although there are some noticeable differences. For example, the combinatorics of finite linear orders is replaced by that of finite rooted trees, which causes some technical complications. Other important differences are that the cartesian product of simplicial sets is replaced by a closed symmetric monoidal structure, which no longer commutes with the nerve construction. Related to this, the category $\Omega$ which takes the place of the simplicial index category $\Delta$ does not have a terminal object, and in fact the homotopy type of its classifying space $B\Omega$ is as yet unknown.[1] Finally, an important gap (at least, up to now) in the theory is the lack of a suitable geometric realization functor which is compatible with this monoidal structure —a functor from dendroidal sets into the category of topological spaces or into a closely related category.

The notes which you have in front of you are a faithful presentation of the lectures that I gave in the context of an advanced course on Simplicial Methods in Higher Categories, as part of a special year on Homotopy Theory and Higher Categories at the CRM in Barcelona. In these lectures, I have reported on recent research, done partly in collaboration with Clemens Berger, Denis-Charles Cisinski and Ittay Weiss, as listed in the references. The lectures owe a lot to their insights, and I apologize for what they undoubtedly consider as an inadequate representation of their ideas about operads and dendroidal sets.

I would like to thank the people involved in the organization of the CRM special year, in particular the coordinators Carles Casacuberta and Joachim Kock, for giving me the opportunity to expose the theory of dendroidal sets, and for providing me with such an inspiring and active audience. But above all, I am immensely grateful to Javier Gutiérrez, who helped me turn scattered notes of the lectures and bits from unpublished papers into a coherent text.

Ieke Moerdijk

---

[1] *Added in proof:* I have shown in the meantime that $B\Omega$ is contractible.

# Lecture 1

# Operads

Operads are tools used to describe algebraic structures in monoidal categories. They are very important in categories with a good notion of homotopy, where they are useful for the study of homotopy invariant algebraic structures and hierarchies of higher homotopies. In this first lecture, we recall the definition of operads and coloured operads, and give some examples.

## 1.1 Operads

In what follows, $\mathcal{E}$ will denote a cocomplete symmetric monoidal category, with tensor product denoted by $\otimes$ and unit $I$. We will also assume that $\mathcal{E}$ is *closed* and write $\mathrm{Hom}_{\mathcal{E}}(X, Y)$ for the internal hom. The symmetric group in $n$ letters will be denoted by $\Sigma_n$.

An *operad* $P$ in $\mathcal{E}$ consists of objects $P(n)$ of $\mathcal{E}$ for $n \geq 0$ together with the following data:

- A *unit*, given by a morphism $I \longrightarrow P(1)$.

- A *composition product* or *substitution*, given by morphisms

$$P(n) \otimes P(k_1) \otimes \cdots \otimes P(k_n) \longrightarrow P(k)$$

  for every $n$ and $k_1, \ldots, k_n$, and $k = \sum_{i=1}^{n} k_i$.

- A *permutation of variables*, given by an action of $\Sigma_n$ from the right on $P(n)$ for every $n$. ($\Sigma_0$ and $\Sigma_1$ will both denote the trivial group.)

One can think of $P(n)$ as the 'space of operations of $n$ variables'. The composition product satisfies well-known equivariance, associativity and unit conditions.

A map of operads $f \colon P \longrightarrow Q$ is given by morphisms $f_n \colon P(n) \longrightarrow Q(n)$ for every $n$ that are compatible with the composition product, the unit and the action of the symmetric group.

I. Moerdijk and B. Toën, *Simplicial Methods for Operads and Algebraic Geometry*, Advanced Courses in Mathematics - CRM Barcelona, DOI 10.1007/978-3-0348-0052-5_1, © Springer Basel AG 2010

For emphasis, we sometimes speak of these operads as *symmetric operads*. A *non-symmetric* or *non-$\Sigma$ operad* is one without the permutations.

An *algebra* for an operad $P$, or a $P$-*algebra*, is an object $A$ of $\mathcal{E}$ together with an action by operations

$$P(n) \otimes A \otimes \overset{(n)}{\cdots} \otimes A \longrightarrow A$$

compatible with the composition product, the unit and the symmetric group actions.

*Example 1.1.1.* The commutative operad $\mathfrak{Com}$ is defined by $\mathfrak{Com}(n) = I$ for all $n$. A $\mathfrak{Com}$-algebra is a commutative monoid in $\mathcal{E}$. The associative operad $\mathcal{Ass}$ is defined by $\mathcal{Ass}(n) = I[\Sigma_n]$, where $I[\Sigma_n]$ denotes a coproduct of copies of the unit of the monoidal category indexed by the elements of $\Sigma_n$. The composition product is described by substitution of block matrices. An $\mathcal{Ass}$-algebra is a monoid in $\mathcal{E}$.

*Example 1.1.2.* Let $\mathcal{E}$ be the category *Top* of topological spaces. Let $D^2 = B(0, 1)$ be the unit disk in the plane with center in the origin. We denote by $C_n(D^2)$ the configuration space of $n$ different points in $D^2$, i.e.,

$$C_n(D^2) = \{(x_1, \ldots, x_n) \mid x_i \in D^2,\ x_i \neq x_j \text{ if } i \neq j\}.$$

The *little 2-disks operad* $\mathcal{D}_2$ is defined as follows. For every $n$, the topological space $\mathcal{D}_2(n)$ is the subspace of $C_n(D^2) \times (0, 1]^n$ consisting of those points $(x_1, \ldots, x_n; t_1, \ldots, t_n)$ such that the disks $B(x_i, t_i)$ are contained in $D^2$ and have disjoint interiors. The following picture represents a point in $\mathcal{D}_2(4)$:

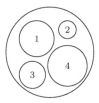

The symmetric group $\Sigma_n$ acts on $\mathcal{D}_2(n)$ by permuting the labels of the disks. The composition product $\gamma$ is defined by composition of embeddings as illustrated in the following example:

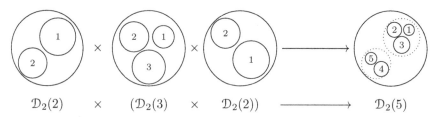

$$\mathcal{D}_2(2) \quad \times \quad (\mathcal{D}_2(3) \quad \times \quad \mathcal{D}_2(2)) \quad \longrightarrow \quad \mathcal{D}_2(5)$$

Any double loop space $\Omega^2 X$ is naturally an algebra over $\mathcal{D}_2$. In fact, by observing that the sphere $S^2$ is the unit disk with its boundary collapsed to a point, one can see that a configuration of $n$ little 2-disks shows how to glue together $n$ based maps $S^2 \longrightarrow X$ to make a single map $S^2 \longrightarrow X$. The converse is essentially true, namely any connected space $X$ that is a $\mathcal{D}_2$-algebra has the homotopy type of a double loop space (see [May72, BV73]).

Every non-$\Sigma$ operad $P$ 'generates' a symmetric operad $\Sigma P$ with the same algebras. The operad $\Sigma P$ is defined by $(\Sigma P)(n) = P(n) \otimes I[\Sigma_n]$. For example, the non-$\Sigma$ operad $P(n) = I$ for every $n$ has as algebras the monoids and $\mathcal{A}ss$ as its associated symmetric operad.

## 1.2 Coloured operads

Let $C$ be a set, whose elements will be called *colours*. A *C-coloured operad* in $\mathcal{E}$ consists of the following data:

- For each sequence $c_1, \ldots, c_n, c$ of elements of $C$, an object $P(c_1, \ldots, c_n; c)$ of $\mathcal{E}$. This object represents the set of operations which take inputs of colours $c_1, \ldots, c_n$ and produce an output of colour $c$.

- *Units*, given by a morphism $I \longrightarrow P(c; c)$ for every $c$ in $C$.

- For every $(n+1)$-tuple of colours $(c_1, \ldots, c_n; c)$ and $n$ given tuples

$$(d_{1,1}, \ldots, d_{1,k_1}; c_1), \ldots, (d_{n,1}, \ldots, d_{n,k_n}; c_n),$$

  a *composition product* given by morphisms

$$P(c_1, \ldots, c_n; c) \otimes P(d_{1,1}, \ldots, d_{1,k_1}; c_1) \otimes \cdots \otimes P(d_{n,1}, \ldots, d_{n,k_n}; c_n)$$
$$\longrightarrow P(d_{1,1}, \ldots, d_{1,k_1}, \ldots, d_{n,1}, \ldots, d_{n,k_n}; c).$$

- *Permutations*, given by an action of the symmetric group. Any $\sigma$ of $\Sigma_n$ gives a map

$$\sigma^* \colon P(c_1, \ldots, c_n; c) \longrightarrow P(c_{\sigma(1)}, \ldots, c_{\sigma(n)}; c).$$

Moreover, the composition product has to be compatible with the action of the symmetric groups and subject to associativity and unitary compatibility relations.

A map of coloured operads from a $C$-coloured operad $P$ to a $D$-coloured operad $Q$ is given by a map $f \colon C \longrightarrow D$ of colours and maps

$$\varphi_{c_1, \ldots, c_n; c} \colon P(c_1, \ldots, c_n; c) \longrightarrow Q(f(c_1), \ldots, f(c_n); f(c))$$

compatible with all the structure maps. We denote by $Oper(\mathcal{E})$ the category whose objects are coloured operads in $\mathcal{E}$ and whose morphisms are maps of coloured operads. We use the notation $Oper$ for the category of coloured operads in the category $Sets$ of sets and functions.

A *P-algebra* now consists of a family of objects $(A_c)_{c \in C}$ of $\mathcal{E}$ together with actions

$$P(c_1, \ldots, c_n; c) \otimes A_{c_1} \otimes \cdots \otimes A_{c_n} \longrightarrow A_c.$$

There is a corresponding notion of non-$\Sigma$ coloured operad (and an associated notion of algebra). Every non-$\Sigma$ $C$-coloured operad $P$ generates a (symmetric) $C$-coloured operad $\Sigma P$ with the same algebras. The operad $\Sigma P$ is defined as follows:

$$(\Sigma P)(c_1, \ldots, c_n; c) = \coprod_{\sigma \in \Sigma_n} P(c_{\sigma^{-1}(1)}, \ldots, c_{\sigma^{-1}(n)}; c). \tag{1.1}$$

In fact, the functor that assigns to every non-$\Sigma$ operad $P$ the symmetric operad $\Sigma P$ is left adjoint to the forgetful functor from the category of (symmetric) $C$-coloured operads to the category of non-$\Sigma$ $C$-coloured operads.

## 1.3   Examples of coloured operads

*Example* 1.3.1. If $C = \{c\}$, then a $C$-coloured operad $P$ is just an ordinary operad, where one writes $P(n)$ instead of $P(c, \ldots, c; c)$ with $n$ inputs.

*Example* 1.3.2. Let $\mathrm{Mod}$ be a coloured operad with two colours $C = \{r, m\}$, for which the only non-zero terms are

$$\mathrm{Mod}(r, \overset{(n)}{\ldots}, r; r) = I[\Sigma_n]$$

for $n \geq 0$, and

$$\mathrm{Mod}_P(c_1, \ldots, c_n; m) = I[\Sigma_n]$$

for $n \geq 1$ when exactly one $c_i$ is $m$ and the rest (if any) are equal to $r$. Then an algebra over $\mathrm{Mod}$ is a pair $(R, M)$ of objects of $\mathcal{E}$ where $R$ is a monoid and $M$ is an object of $\mathcal{E}$ on which $R$ acts (i.e., a module over $R$).

*Example* 1.3.3. Every small category $\mathcal{C}$ can be viewed as a coloured operad in *Sets*. The set of colours is the set of objects of $\mathcal{C}$ and the only operations are unary operations. The composition product is given by the composition between sets of morphisms in $\mathcal{C}$. The algebras over this coloured operad are the covariant functors from $\mathcal{C}$ to *Sets*. Similarly, any small $\mathcal{E}$-enriched category can be viewed as an operad in $\mathcal{E}$.

Conversely, the unary operations in a $C$-coloured operad $P$ give an $\mathcal{E}$-enriched category whose objects are the elements of $C$. These two constructions define adjoint functors

$$j_! : Cat(\mathcal{E}) \rightleftarrows Oper(\mathcal{E}) : j^*,$$

where $Cat(\mathcal{E})$ denotes the category of small $\mathcal{E}$-enriched categories.

*Example* 1.3.4. Every monoidal category $\mathcal{E}$ can be viewed as a coloured operad $\underline{\mathcal{E}}$ in *Sets*. The set of colours is the set of objects of $\mathcal{E}$, and

$$\underline{\mathcal{E}}(E_1, \ldots, E_n; E) = \mathcal{E}(E_1 \otimes \cdots \otimes E_n; E).$$

An $\underline{\mathcal{E}}$-algebra $X$ consists of a set $X(E)$ for every object $E$ of $\mathcal{E}$ and a map

$$X(f)\colon X(E_1) \times \cdots \times X(E_n) \longrightarrow X(E)$$

for every map $f\colon E_1 \otimes \cdots \otimes E_n \longrightarrow E$ in $\mathcal{E}$. In particular, for $n = 2$ and $n = 0$ we obtain maps $X(E) \times X(F) \longrightarrow X(E \otimes F)$ and $* \longrightarrow X(I)$. Hence $\underline{\mathcal{E}}$-algebras are lax monoidal functors from $\mathcal{E}$ to *Sets*.

More generally, any $\mathcal{V}$-enriched monoidal category $\mathcal{E}$ can be viewed as an operad in $\mathcal{V}$. Its algebras are lax monoidal $\mathcal{V}$-functors from $\mathcal{E}$ to $\mathcal{V}$.

*Example* 1.3.5. For a given set $S$, there is a non-$\Sigma$ operad $\mathcal{C}at_S$ in *Sets* whose algebras are categories with the set $S$ as objects. The set of colours for this operad is $S \times S$ and

$$\mathcal{C}at_S((s_1, s_1'), \ldots, (s_n, s_n'); (s, s')) = \begin{cases} * & \text{if } s_1' = s_2, s_2' = s_3, \ldots, s_1 = s, s_n' = s', \\ \emptyset & \text{otherwise;} \end{cases}$$

$$\mathcal{C}at_S(\ ; (s, s')) = \begin{cases} * & \text{if } s = s', \\ \emptyset & \text{otherwise.} \end{cases}$$

For a $\mathcal{C}at_S$-algebra $A$, the set $A_{(s,s')}$ is the set of arrows from $s$ to $s'$.

*Example* 1.3.6. Let $C$ be a set of colours and $\mathcal{E}$ any closed symmetric monoidal category. We describe a coloured operad $S_{\mathcal{E}}^C$ in $\mathcal{E}$ whose algebras are $C$-coloured operads in $\mathcal{E}$. Observe that it is enough to describe an operad $S^C$ in the category of sets with this property. In fact, then the strong symmetric monoidal functor $(-)_{\mathcal{E}}\colon Sets \longrightarrow \mathcal{E}$ defined as $X_{\mathcal{E}} = \coprod_{x \in X} I$ sends coloured operads to coloured operads. Hence the image of $S^C$ under this functor, denoted by $S_{\mathcal{E}}^C$, is a coloured operad in $\mathcal{E}$ whose algebras are $C$-coloured operads in $\mathcal{E}$.

The colours of $S^C$ are

$$\mathrm{col}(S^C) = \{(c_1, \ldots, c_n; c) \mid c_i, c \in C, n \geq 0\}.$$

We will use the following notation: $\bar{c}_i = (c_{i,1}, \ldots, c_{i,k_i}; c_i)$ and $\bar{a} = (a_1, \ldots, a_m; a)$. For each $(n + 1)$-tuple of colours $(\bar{c}_1, \ldots, \bar{c}_n; \bar{a})$ the elements of $S^C(\bar{c}_1, \ldots, \bar{c}_n; \bar{a})$ are equivalence classes of triples $(T, \sigma, \tau)$ where:

- $T$ is a planar rooted tree with $m$ input edges coloured by $a_1, \ldots, a_m$, a root edge coloured by $a$, and $n$ vertices. See §2.1 for an explanation of the terminology.

- $\sigma$ is a bijection $\sigma\colon \{1, \ldots, n\} \longrightarrow V(T)$ with the property that $\sigma(i)$ has $k_i$ input edges coloured from left to right by $c_{i,1}, \ldots, c_{i,k_i}$ and one output edge coloured by $c_i$.

- $\tau$ is a bijection $\tau\colon \{1, \ldots, m\} \longrightarrow \mathrm{in}(T)$ such that $\tau(i)$ has colour $a_i$. Here $\mathrm{in}(T)$ denotes the set of input edges of $T$.

Two such triples $(T, \sigma, \tau)$, $(T'\sigma', \tau')$ are equivalent if and only if there is a planar isomorphism $\varphi \colon T \longrightarrow T'$ such that $\varphi \circ \sigma = \sigma'$, $\varphi \circ \tau = \tau'$, and $\varphi$ respects the colouring, i.e., if $e$ is an edge of $T$ of colour $c$, then the edge $\varphi(e)$ in $T'$ has colour $c$ too. Any element $\alpha$ in $\Sigma_n$ induces a map

$$\alpha^* \colon S^C(\bar{c}_1, \dots, \bar{c}_n; \bar{a}) \longrightarrow S^C(\bar{c}_{\alpha(1)}, \dots, \bar{c}_{\alpha(n)}; \bar{a})$$

that sends $(T, \sigma, \tau)$ to $(T, \sigma \circ \alpha, \tau)$. That is, $\alpha^*(T)$ is the same tree as $T$ but with a renumbering of the vertices given by $\alpha$.

Let $\sigma$ be any element in $\Sigma_m$ and $\bar{a}_1 = (a_{\sigma(1)}, \dots, a_{\sigma(m)}; a)$. The set $S^C(\bar{a}_1; \bar{a})$ can be identified with the subset of elements of $\Sigma_m$ that permute $(a_{\sigma(1)}, \dots, a_{\sigma(m)})$ into $(a_1, \dots, a_m)$. In particular, if $\bar{a} = \bar{a}_1$ then the set $S^C(\bar{a}; \bar{a})$ can be identified with the (opposite) subgroup of $\Sigma_m$ that leaves the colours $a_1, \dots, a_m$ invariant. There is a distinguished element $1_{\bar{a}}$ in $S^C(\bar{a}; \bar{a})$ corresponding to the tree

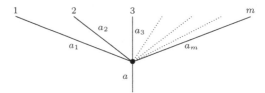

The composition product on $S^C$ is defined as follows. Given an element $(T, \sigma, \tau)$ of $S^C(\bar{c}_1, \dots, \bar{c}_n; \bar{a})$ and $n$ elements $(T_1, \sigma_1, \tau_1), \dots, (T_n, \sigma_n, \tau_n)$ of

$$S^C(\bar{d}_{1,1}, \dots, \bar{d}_{1,k_1}; \bar{c}_1), \dots, S^C(\bar{d}_{n,1}, \dots, \bar{d}_{n,k_n}; \bar{c}_n)$$

respectively, we get an element $T'$ of

$$S^C(\bar{d}_{1,1}, \dots, \bar{d}_{1,k_1}, \bar{d}_{2,1}, \dots, \bar{d}_{2,k_2}, \dots, \bar{d}_{n,1}, \dots, \bar{d}_{n,k_n}; \bar{a})$$

in the following way:

(i) $T'$ is obtained by replacing the vertex $\sigma(i)$ of $T$ by the tree $T_i$ for every $i$. This is done by identifying the input edges of $\sigma(i)$ in $T$ with the input edges of $T_i$ via the bijection $\tau_i$. The $c_{i,j}$-coloured input edge of $\sigma(i)$ is matched with the $c_{i,j}$-coloured input edge $\tau_i(j)$ of $T_i$. (Note that the colours of the input edges and the output of $\sigma(i)$ coincide with the colours of the input edges and the root of $T_i$.)

(ii) The vertices of $T'$ are numbered following the order, i.e., first number the subtree $T_1$ in $T'$ ordered by $\sigma_1$, then $T_2$ ordered by $\sigma_2$, and so on.

(iii) The input edges of $T'$ are numbered following $\tau$ and the identifications given by $\tau_i$.

The above composition product endows the collection $S^C$ with a coloured operad structure. Algebras over this operad $S^C$ are precisely $C$-coloured operads in *Sets*.

# Lecture 2

# Trees as operads

In this lecture, we introduce convenient categories of trees that will be used for the definition of dendroidal sets. These categories are generalizations of the simplicial category $\Delta$ used to define simplicial sets. First we consider the case of planar trees and then the more general case of non-planar trees.

## 2.1   A formalism of trees

A *tree* is a non-empty connected finite graph with no loops. A vertex in a graph is called *outer* if it has only one edge attached to it. All the trees we will consider are *rooted trees*, i.e., equipped with a distinguished outer vertex called the *output* and a (possibly empty) set of outer vertices (not containing the output vertex) called the set of *inputs*.

When drawing trees, we will delete the output and input vertices from the picture. From now on, the term 'vertex' in a tree will always refer to a remaining vertex. Given a tree $T$, we denote by $V(T)$ the set of vertices of $T$ and by $E(T)$ the set of edges of $T$.

The edges attached to the deleted input vertices are called *input edges* or *leaves*; the edge attached to the deleted output vertex is called *output edge* or *root*. The rest of the edges are called *inner edges*. The root induces an obvious direction in the tree, 'from the leaves towards the root'. If $v$ is a vertex of a finite rooted tree, we denote by out($v$) the unique outgoing edge and by in($v$) the set of incoming edges (note that in($v$) can be empty). The cardinality of in($v$) is called the *valence* of $v$, the element of out($v$) is the *output* of $v$, and the elements of in($v$) are the *inputs* of $v$.

As an example, consider the following picture of a tree:

I. Moerdijk and B. Toën, *Simplicial Methods for Operads and Algebraic Geometry*, Advanced Courses in Mathematics - CRM Barcelona, DOI 10.1007/978-3-0348-0052-5_2, © Springer Basel AG 2010

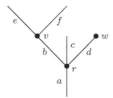

The output vertex at the edge $a$ and the input vertices at $e$, $f$ and $c$ have been deleted. This tree has three vertices $r$, $v$ and $w$ of respective valences 3, 2, and 0. It also has three input edges or leaves, namely $e$, $f$ and $c$. The edges $b$ and $d$ are inner edges and the edge $a$ is the root. A tree with no vertices

whose input edge (which we denote by $e$) coincides with its output edge will be denoted by $\eta_e$, or simply by $\eta$.

**Definition 2.1.1.** A *planar rooted tree* is a rooted tree $T$ together with a linear ordering of in$(v)$ for each vertex $v$ of $T$.

The ordering of in$(v)$ for each vertex is equivalent to drawing the tree on the plane. When we draw a tree we will always put the root at the bottom. One drawback of drawing a tree on the plane is that it immediately becomes a planar tree; we thus may have many different 'pictures' for the same tree. For example, the two trees

are two different planar representations of the same tree.

## 2.2   Planar trees

Let $T$ be a planar rooted tree. Any such tree generates a non-$\Sigma$ operad, which we denote by $\Omega_p(T)$. The set of colours of $\Omega_p(T)$ is the set $E(T)$ of edges of $T$, and the operations are generated by the vertices of the tree. More explicitly, each vertex $v$ with input edges $e_1, \ldots, e_n$ and output edge $e$ defines an operation $v \in \Omega_p(T)(e_1, \ldots, e_n; e)$. The other operations are the unit operations and the operations obtained by compositions. This operad has the property that, for all $e_1, \ldots, e_n, e$, the set of operations $\Omega_p(T)(e_1, \ldots, e_n; e)$ contains at most one element. For example, consider the same tree $T$ pictured before:

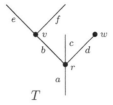

$T$

The operad $\Omega_p(T)$ has six colours $a$, $b$, $c$, $d$, $e$, and $f$. Then $v \in \Omega_p(T)(e, f; b)$, $w \in \Omega_p(T)(\,; d)$, and $r \in \Omega_p(b, c, d; a)$ are the generators, while the other operations are the units $1_a, 1_b, \ldots, 1_f$ and the operations obtained by compositions, namely $r \circ_1 v \in \Omega_p(T)(e, f, c, d; a)$, $r \circ_3 w \in \Omega_p(T)(b, c; a)$, and

$$r(v, 1_c, w) = (r \circ_1 v) \circ_4 w = (r \circ_3 w) \circ_1 v \in \Omega_p(T)(e, f, c; a).$$

This is a complete description of the operad $\Omega_p(T)$.

**Definition 2.2.1.** The *category of planar rooted trees* $\Omega_p$ is the full subcategory of the category of non-$\Sigma$ coloured operads whose objects are $\Omega_p(T)$ for any tree $T$.

We can view $\Omega_p$ as the category whose objects are planar rooted trees. The set of morphisms from a tree $S$ to a tree $T$ is given by the set of non-$\Sigma$ coloured operad maps from $\Omega_p(S)$ to $\Omega_p(T)$. Observe that any morphism $S \longrightarrow T$ in $\Omega_p$ is completely determined by its effect on the colours (i.e., edges).

The category $\Omega_p$ extends the simplicial category $\Delta$. Indeed, any $n \geq 0$ defines a linear tree

$L_n$

with $n + 1$ edges and $n$ vertices $v_1, \ldots, v_n$. We denote this tree by $[n]$ or $L_n$. Any order-preserving map $\{0, \ldots, n\} \longrightarrow \{0, \ldots, m\}$ defines an arrow $[n] \longrightarrow [m]$ in the category $\Omega_p$. In this way, we obtain an embedding

$$\Delta \xrightarrow{\ u\ } \Omega_p.$$

This embedding is fully faithful. Moreover, it describes $\Delta$ as a sieve (or ideal) in $\Omega_p$, in the sense that for any arrow $S \longrightarrow T$ in $\Omega_p$, if $T$ is linear then so is $S$. In the next sections we give a more explicit description of the morphisms in $\Omega_p$.

## 2.2.1  Face maps

Let $T$ be a planar rooted tree and $b$ an inner edge in $T$. Let us denote by $T/b$ the tree obtained from $T$ by contracting $b$. Then there is a natural map $\partial_b \colon T/b \longrightarrow T$ in $\Omega_p$, called the *inner face* map associated with $b$. This map is the inclusion on both the colours and the generating operations of $\Omega_p(T/b)$, except for the operation $u$, which is sent to $r \circ_b v$. Here $r$ and $v$ are the two vertices in $T$ at the two ends of $b$, and $u$ is the corresponding vertex in $T/b$, as in the picture:

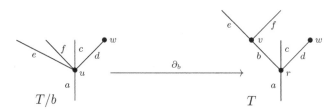

Now let $T$ be a planar rooted tree and $v$ a vertex of $T$ with exactly one inner edge attached to it. Let $T/v$ be the tree obtained from $T$ by removing the vertex $v$ and all the outer edges. There is a face map associated to this operation, denoted $\partial_v \colon T/v \longrightarrow T$, which is the inclusion both on the colours and on the generating operations of $\Omega_p(T/v)$. These types of face maps are called the *outer faces* of $T$. The following are two outer face maps:

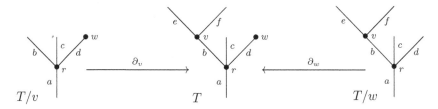

Note that the possibility of removing the root vertex of $T$ is included in this definition. This situation can happen only if the root vertex is attached to exactly one inner edge, thus not every tree $T$ has an outer face induced by its root. There is another particular situation which requires special attention, namely the inclusion of the tree with no vertices $\eta$ into a tree with one vertex, called a *corolla*. In this case we get $n+1$ face maps if the corolla has $n$ leaves. The operad $\Omega_p(\eta)$ consists of only one colour and the identity operation on it. Then a map of operads $\Omega_p(\eta) \longrightarrow \Omega_p(T)$ is just a choice of an edge of $T$.

We will use the term *face map* to refer to an inner or outer face map.

## 2.2.2  Degeneracy maps

There is one more type of map that can be associated with a vertex $v$ of valence one in $T$ as follows. Let $T\backslash v$ be the tree obtained from $T$ by removing the vertex $v$ and merging the two edges incident to it into one edge $e$. Then there is a map

$\sigma_v : T \longrightarrow T\backslash v$ in $\Omega_p$ called the *degeneracy map* associated with $v$, which sends the colours $e_1$ and $e_2$ of $\Omega_p(T)$ to $e$, sends the generating operation $v$ to $\mathrm{id}_e$, and is the identity for the other colours and operations. It can be pictured like this:

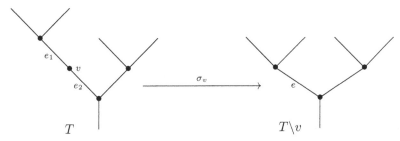

Face maps and degeneracy maps generate the whole category $\Omega_p$. The following lemma is the generalization to $\Omega_p$ of the well-known fact that in $\Delta$ each arrow can be written as a composition of degeneracy maps followed by face maps. For the proof of this fact we refer the reader to Lemma 2.3.2, where we prove a similar statement in the category of non-planar trees.

**Lemma 2.2.2.** *Any arrow $f : A \longrightarrow B$ in $\Omega_p$ decomposes (up to isomorphism) as*

$$
\begin{array}{ccc}
A & \xrightarrow{\;f\;} & B \\
 & \searrow{\scriptstyle \sigma} & \big\uparrow{\scriptstyle \delta} \\
 & & C
\end{array}
$$

*where $\sigma : A \longrightarrow C$ is a composition of degeneracy maps and $\delta : C \longrightarrow B$ is a composition of face maps.* $\qquad\square$

### 2.2.3 Dendroidal identities

In this section we are going to make explicit the relations between the generating maps (faces and degeneracies) of $\Omega_p$. The identities that we obtain generalize the simplicial ones in the category $\Delta$.

**Elementary face relations**

Let $\partial_a : T/a \longrightarrow T$ and $\partial_b : T/b \longrightarrow T$ be distinct inner faces of $T$. It follows that the inner faces $\partial_a : (T/b)/a \longrightarrow T/b$ and $\partial_b : (T/a)/b \longrightarrow T/a$ exist, we have $(T/a)/b = (T/b)/a$, and the following diagram commutes:

$$
\begin{array}{ccc}
(T/a)/b & \xrightarrow{\;\partial_b\;} & T/a \\
{\scriptstyle \partial_a}\big\downarrow & & \big\downarrow{\scriptstyle \partial_a} \\
T/b & \xrightarrow{\;\partial_b\;} & T.
\end{array}
$$

Let $\partial_v : T/v \longrightarrow T$ and $\partial_w : T/w \longrightarrow T$ be distinct outer faces of $T$, and assume that $T$ has at least three vertices. Then the outer faces $\partial_w : (T/v)/w \longrightarrow T/v$ and $\partial_v : (T/w)/v \longrightarrow T/w$ also exist, $(T/v)/w = (T/w)/v$, and the following diagram commutes:

$$
\begin{array}{ccc}
(T/v)/w & \xrightarrow{\ \partial_w\ } & T/v \\
{\scriptstyle\partial_v}\downarrow & & \downarrow{\scriptstyle\partial_v} \\
T/w & \xrightarrow{\ \partial_w\ } & T.
\end{array}
$$

In case that $T$ has only two vertices, there is a similar commutative diagram involving the inclusion of $\eta$ into the $n$-th corolla.

The last remaining case is when we compose an inner face with an outer one in any order. There are several possibilities and in all of them we suppose that $\partial_v : T/v \longrightarrow T$ is an outer face and $\partial_e : T/e \longrightarrow T$ is an inner face.

- If in $T$ the edge $e$ is not adjacent to the vertex $v$, then the outer face $\partial_v : (T/e)/v \longrightarrow T/e$ and the inner face $\partial_e : (T/v)/e \longrightarrow T/v$ exist, $(T/e)/v = (T/v)/e$, and the following diagram commutes:

$$
\begin{array}{ccc}
(T/v)/e & \xrightarrow{\ \partial_e\ } & T/v \\
{\scriptstyle\partial_v}\downarrow & & \downarrow{\scriptstyle\partial_v} \\
T/e & \xrightarrow{\ \partial_e\ } & T.
\end{array}
$$

- Suppose that in $T$ the inner edge $e$ is adjacent to the vertex $v$ and denote the other adjacent vertex to $e$ by $w$. Observe that $v$ and $w$ contribute a vertex $v \circ_e w$ or $w \circ_e v$ to $T/e$. Let us denote this vertex by $z$. Then the outer face $\partial_z : (T/e)/z \longrightarrow T/e$ exists if and only if the outer face $\partial_w : (T/v)/w \longrightarrow T/v$ exists, and in this case $(T/e)/z = (T/v)/w$. Moreover, the following diagram commutes:

$$
\begin{array}{ccccc}
(T/v)/w & = & (T/e)/z & \xrightarrow{\ \partial_z\ } & T/e \\
{\scriptstyle\partial_w}\downarrow & & & & \downarrow{\scriptstyle\partial_e} \\
T/v & & \xrightarrow{\qquad\qquad\partial_v\qquad\qquad} & & T.
\end{array}
$$

It follows that we can write $\partial_v\partial_w = \partial_e\partial_z$, where $z = v \circ_e w$ if $v$ is 'closer' to the root of $T$ or $z = w \circ_e v$ if $w$ is 'closer' to the root of $T$.

## Elementary degeneracy relations

Let $\sigma_v : T \longrightarrow T\backslash v$ and $\sigma_w : T \longrightarrow T\backslash w$ be two degeneracies of $T$. Then the degeneracies $\sigma_v : T\backslash w \longrightarrow (T\backslash w)\backslash v$ and $\sigma_w : T\backslash v \longrightarrow (T\backslash v)\backslash w$ exist, we have

$(T\backslash v)\backslash w = (T\backslash w)\backslash v$, and the following diagram commutes:

$$\begin{array}{ccc} T & \xrightarrow{\ \sigma_v\ } & T\backslash v \\ {\scriptstyle \sigma_w}\downarrow & & \downarrow{\scriptstyle \sigma_w} \\ T\backslash w & \xrightarrow{\ \sigma_v\ } & (T\backslash v)\backslash w. \end{array}$$

**Combined relations**

Let $\sigma_v\colon T \longrightarrow T\backslash v$ be a degeneracy and $\partial\colon T' \longrightarrow T$ be a face map such that $\sigma_v\colon T' \longrightarrow T'\backslash v$ makes sense (i.e., $T'$ still contains $v$ and its two adjacent edges as a subtree). Then there exists an induced face map $\partial\colon T'\backslash v \longrightarrow T\backslash v$ determined by the same vertex or edge as $\partial\colon T' \longrightarrow T$. Moreover, the following diagram commutes:

$$\begin{array}{ccc} T & \xrightarrow{\ \sigma_v\ } & T\backslash v \\ {\scriptstyle \partial}\uparrow & & \uparrow{\scriptstyle \partial} \\ T' & \xrightarrow{\ \sigma_v\ } & T'\backslash v. \end{array}$$

Let $\sigma_v\colon T \longrightarrow T\backslash v$ be a degeneracy and $\partial\colon T' \longrightarrow T$ be a face map induced by one of the adjacent edges to $v$ or the removal of $v$, if that is possible. It follows that $T' = T\backslash v$ and the composition

$$T\backslash v \xrightarrow{\ \partial\ } T \xrightarrow{\ \sigma_v\ } T\backslash v$$

is the identity map $\mathrm{id}_{T\backslash v}$.

## 2.3 Non-planar trees

Any non-planar tree $T$ generates a (symmetric) coloured operad $\Omega(T)$. Similarly as in the case of planar trees, the set of colours of $\Omega(T)$ is the set of edges $E(T)$ of $T$. The operations are generated by the vertices of the tree, and the symmetric group on $n$ letters $\Sigma_n$ acts on each operation with $n$ inputs by permuting the order of its inputs. Each vertex $v$ of the tree with output edge $e$ and a numbering of its input edges $e_1, \ldots, e_n$ defines an operation $v \in \Omega(e_1, \ldots, e_n; e)$. The other operations are the unit operations and the operations obtained by compositions and the action of the symmetric group. For example, consider the tree

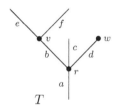

The operad $\Omega(T)$ has six colours $a$, $b$, $c$, $d$, $e$, and $f$. The generating operations are the same as the generating operations of $\Omega_p(T)$. All the operations of $\Omega_p(T)$ are operations of $\Omega(T)$, but there are more operations in $\Omega(T)$ obtained by the action of the symmetric group. For example if $\sigma$ is the transposition of two elements of $\Sigma_2$, we have an operation $v \circ \sigma \in \Omega(f, e; b)$. Similarly if $\sigma$ is the transposition of $\Sigma_3$ that interchanges the first and third elements, then there is an operation $r \circ \sigma \in \Omega(d, c, b; a)$.

More formally, if $T$ is any tree, then $\Omega(T) = \Sigma(\Omega_p(\overline{T}))$, where $\overline{T}$ is a planar representative of $T$. In fact, a choice of a planar structure on $T$ is precisely a choice of generators for $\Omega(T)$.

**Definition 2.3.1.** The *category of rooted trees* $\Omega$ is the full subcategory of the category of coloured operads whose objects are $\Omega(T)$ for any tree $T$.

We can view $\Omega$ as the category whose objects are rooted trees. The set of morphisms from a tree $S$ to a tree $T$ is given by the set of coloured operad maps from $\Omega(S)$ to $\Omega(T)$. Note that any morphism $S \longrightarrow T$ in $\Omega$ is completely determined by its effect on the colours (i.e., edges).

The morphisms in $\Omega$ are generated by faces and degeneracies (as in the planar case) and also by (non-planar) isomorphisms.

**Lemma 2.3.2.** *Any arrow* $f: S \longrightarrow T$ *in* $\Omega$ *decomposes as*

$$
\begin{array}{ccc}
S & \xrightarrow{f} & T \\
\downarrow{\scriptstyle \sigma} & & \uparrow{\scriptstyle \delta} \\
S' & \xrightarrow{\varphi} & T'
\end{array}
$$

*where* $\sigma: S \longrightarrow S'$ *is a composition of degeneracy maps,* $\varphi: S' \longrightarrow T'$ *is an isomorphism, and* $\delta: T' \longrightarrow T$ *is a composition of face maps.*

*Proof.* We proceed by induction on the sum of the number of vertices of $S$ and $T$. If $T$ and $S$ have no vertices, then $T = S = \eta$ and $f$ is the identity. Note that, without loss of generality, we can assume that $f$ sends the root of $S$ to the root of $T$; otherwise we can factor it as a map $S \longrightarrow T'$ that preserves the root followed by a map $T' \longrightarrow T$ that is a composition of outer faces. Also, we can assume that $f$ is an epimorphism on the leaves since, if this is not the case, $f$ factors as $S \longrightarrow T/v \xrightarrow{\partial_v} T$, where $v$ is the vertex below the leaf in $T$ that is not in the image of $f$.

If $a$ and $b$ are edges of $S$ such that $f(a) = f(b)$, then $a$ and $b$ must be on the same (linear) branch of $S$ and $f$ sends intermediate vertices to identities.

Since $f$ is a map of coloured operads, we can factor it in a unique way as a surjection followed by an injection on the colours. This corresponds to a factorization in $\Omega$,

$$
S \xrightarrow{\psi} S' \xrightarrow{\xi} T,
$$

where $\psi$ is a composition of degeneracies and $\xi$ is bijective on leaves, sends the root of $S'$ to the root of $T$, and is injective on the colours (by the previous observations).

If $\xi$ is surjective on colours, then $\xi$ is an isomorphism. If $\xi$ is not surjective, then there is an edge $e$ in $T$ not in the image of $\xi$. Since $e$ is an internal edge (not a leaf), $\xi$ factors as

$$S' \xrightarrow{\xi'} T/e \xrightarrow{\partial_e} T.$$

Now we continue by induction on the map $\xi'$. □

In general, limits and colimits do not exist in the category $\Omega$; for example, $\Omega$ lacks sums and products. However, certain pushouts do exist in $\Omega$, as expressed in the following lemma:

**Lemma 2.3.3.** *Let* $f \colon R \twoheadrightarrow S$ *and* $g \colon R \twoheadrightarrow T$ *be two surjective maps in* $\Omega$. *Then the pushout*

$$\begin{array}{ccc} R & \xrightarrow{f} & S \\ {\scriptstyle g}\big\downarrow & & \big\downarrow \\ T & \dashrightarrow & P \end{array}$$

*exists in* $\Omega$.

*Proof.* The maps $f$ and $g$ can each be written as a composition of an isomorphism and a sequence of degeneracy maps by Lemma 2.3.2. Since pushout squares can be pasted together to get larger pushout squares, it thus suffices to prove the lemma in the case where $f$ and $g$ are degeneracy maps given by unary vertices $v$ and $w$ in $R$, i.e., $f \colon R \longrightarrow S$ is $\sigma_v \colon R \longrightarrow R\backslash v$ and $g \colon R \longrightarrow T$ is $\sigma_w \colon R \longrightarrow R\backslash w$. If $v = w$, then the following diagram is a pushout:

$$\begin{array}{ccc} R & \xrightarrow{\sigma_v} & R\backslash v \\ {\scriptstyle \sigma_v}\big\downarrow & & \big\| \\ R\backslash v & =\!\!=\!\!= & R\backslash v. \end{array}$$

If $v \neq w$, then the commutative square

$$\begin{array}{ccc} R & \xrightarrow{\sigma_v} & R\backslash v \\ {\scriptstyle \sigma_w}\big\downarrow & & \big\downarrow{\scriptstyle \sigma_w} \\ R\backslash w & \xrightarrow{\sigma_v} & (R\backslash v)\backslash w = (R\backslash w)\backslash v \end{array}$$

is also a pushout, as one easily checks. □

### 2.3.1 Dendroidal identities with isomorphisms

The dendroidal identities for the category $\Omega$ are the same as for the category $\Omega_p$ plus some more relations involving the isomorphisms in $\Omega$. As an example, we give the following relation, that involves inner faces and isomorphisms. Let $T$ be a tree with an inner edge $a$ and let $f: T \longrightarrow T'$ be a (non-planar) isomorphism. Then the trees $T/a$ and $T'/b$ exist, where $b = f(a)$, the map $f$ restricts to an isomorphism $f: T/a \longrightarrow T'/b$, and the following diagram commutes:

$$
\begin{array}{ccc}
T/a & \xrightarrow{\ \partial_a\ } & T \\
{\scriptstyle f}\downarrow & & \downarrow{\scriptstyle f} \\
T'/b & \xrightarrow{\ \partial_b\ } & T'.
\end{array}
$$

Similar relations hold for outer faces and degeneracies.

### 2.3.2 Isomorphisms along faces and degeneracies

For any tree $T$ in $\Omega$, let $P(T)$ be the set of planar structures of $T$. Note that $P(T) \neq \emptyset$ for every tree $T$. Thus, the category $\Omega$ is equivalent to the category $\Omega'$ whose objects are planar trees, i.e., pairs $(T, p)$ where $T$ is an object of $\Omega$ and $p \in P(T)$, and whose morphisms are given by

$$\Omega'((T, p), (T', p')) = \Omega(T, T').$$

A morphism $\varphi: (T, p) \longrightarrow (T', p')$ in $\Omega'$ is called *planar* if, when we pull back the planar structure $p'$ on $T'$ to one on $T$ along $\varphi$, then it coincides with $p$. Using this equivalent formulation of $\Omega$, the category $\Omega_p$ is then the subcategory of $\Omega$ consisting of the same objects and planar maps only, i.e., compositions of faces and degeneracies. In $\Omega_p$, the only automorphisms are identities.

If $\delta: T \rightarrowtail S$ is a composition of faces and $\alpha: S \longrightarrow S'$ is an isomorphism, there is a factorization

$$
\begin{array}{ccc}
T & \xrightarrow{\ \delta\ } & S \\
{\scriptstyle \alpha'}\downarrow{\scriptstyle \sim} & & {\scriptstyle \sim}\downarrow{\scriptstyle \alpha} \\
T' & \xdashrightarrow{\ \delta'\ } & S',
\end{array}
$$

where $\delta'$ is again a composition of faces and $\alpha'$ is an isomorphism. This factorization is unique if one fixes some conventions, e.g., one takes the objects of $\Omega$ to be *planar* trees, and takes faces and degeneracies to be planar maps. Similarly, isomorphisms can be pushed forward and pulled back along a composition of degeneracies. Let $\sigma: T \longrightarrow S$ be a composition of degeneracies and $\alpha: S \longrightarrow S'$ and

$\beta\colon T \longrightarrow T'$ be two isomorphisms. Then there are factorizations

$$
\begin{array}{ccc}
T & \xrightarrow{\ \sigma\ } & S \\
{\scriptstyle \alpha'}\big\downarrow & & \big\downarrow{\scriptstyle \alpha} \\
T' & \xdashrightarrow{\ \sigma'\ } & S'
\end{array}
\qquad\qquad
\begin{array}{ccc}
T & \xrightarrow{\ \sigma\ } & S \\
{\scriptstyle \beta}\big\downarrow & & \big\downarrow{\scriptstyle \beta'} \\
T' & \xdashrightarrow{\ \sigma''\ } & S'
\end{array}
$$

where $\alpha'$ and $\beta'$ are isomorphisms and $\sigma'$ and $\sigma''$ are compositions of degeneracies.

Thus, any arrow in $\Omega$ can be written in the form $\delta\sigma\alpha$ or $\delta\alpha\sigma$ with $\delta$ a composition of faces, $\sigma$ a composition of degeneracies, and $\alpha$ an isomorphism.

### 2.3.3  The presheaf of planar structures

Let $P\colon \Omega^{\mathrm{op}} \longrightarrow Sets$ be the presheaf on $\Omega$ that sends each tree to its set of planar structures. Observe that $P(T)$ is a torsor under $\mathrm{Aut}(T)$ for every tree $T$, where $\mathrm{Aut}(T)$ denotes the set of automorphisms of $T$. Recall that the category of elements $\Omega/P$ is the category whose objects are pairs $(T, x)$ with $x \in P(T)$. A morphism between two objects $(T, x)$ and $(S, y)$ is given by a morphism $f\colon T \longrightarrow S$ in $\Omega$ such that $P(f)(y) = x$. Hence, $\Omega/P = \Omega_p$ and we have a projection $v\colon \Omega_p \longrightarrow \Omega$. There is a commutative triangle

where $i$ is the fully faithful embedding of $\Delta$ into $\Omega$ which sends the object $[n]$ in $\Delta$ to the linear tree $L_n$ with $n$ vertices and $n + 1$ edges for every $n \geq 0$.

### 2.3.4  Relation with the simplicial category

We have seen that both the categories $\Omega$ and $\Omega_p$ extend the category $\Delta$, by viewing the objects of $\Delta$ as linear trees. In fact, it is possible to obtain $\Delta$ as a comma category of $\Omega$ or of $\Omega_p$ as follows.

Let $\eta$ be the tree in $\Omega$ consisting of no vertices and one edge, and let $\eta_p$ be the planar representative of $\eta$ in $\Omega_p$. If $T$ is any tree in $\Omega$, then $\Omega(T, \eta)$ consists of only one morphism if $T$ is a linear tree, or it is the empty set otherwise. The same happens for $\Omega_p$ and $\eta_p$. Thus, $\Omega/\eta = \Omega_p/\eta_p = \Delta$.

# Lecture 3

# Dendroidal sets

In this lecture, we introduce the basic notions and terminology for the category of dendroidal sets.

## 3.1 Basic definitions and examples

In this section we define the categories of dendroidal sets and planar dendroidal sets as categories of presheaves over $\Omega$ and $\Omega_p$. We establish the relation of these categories with the category of simplicial sets and with the category of operads by means of natural adjoint functors between them. Namely, we construct a dendroidal nerve functor from operads to dendroidal sets generalizing the classical nerve construction from small categories to simplicial sets.

**Definition 3.1.1.** The category *dSets* of *dendroidal sets* is the category of presheaves on $\Omega$. The objects are functors $\Omega^{\mathrm{op}} \longrightarrow Sets$ and the morphisms are given by natural transformations. The category *pdSets* of *planar dendroidal sets* is defined similarly by replacing $\Omega$ by $\Omega_p$.

Thus, a dendroidal set $X$ is given by a set $X(T)$, denoted by $X_T$, for each tree $T$, together with a map $\alpha^* \colon X_T \longrightarrow X_S$ for each morphism $\alpha \colon S \longrightarrow T$ in $\Omega$. Since $X$ is a functor, $(\mathrm{id})^* = \mathrm{id}$, and if $\alpha \colon S \longrightarrow T$ and $\beta \colon R \longrightarrow S$ are morphisms in $\Omega$, then $(\alpha \circ \beta)^* = \beta^* \circ \alpha^*$. The set $X_T$ is called the set of *dendrices of shape $T$* or the set of *$T$-dendrices*.

A *morphism of dendroidal sets* $f \colon X \longrightarrow Y$ is given by maps $f \colon X_T \longrightarrow Y_T$ for each tree $T$, commuting with the structure maps, i.e., if $\alpha \colon S \longrightarrow T$ is any morphism in $\Omega$ and $x \in X_T$, then $f(\alpha^* x) = \alpha^* f(x)$.

Given two dendroidal sets $Y$ and $X$, we say that $Y$ is a *dendroidal subset* of $X$ if $Y_T \subseteq X_T$ for every tree $T$ and the inclusion map $Y \hookrightarrow X$ is a morphism of dendroidal sets.

I. Moerdijk and B. Toën, *Simplicial Methods for Operads and Algebraic Geometry*, Advanced Courses in Mathematics - CRM Barcelona, DOI 10.1007/978-3-0348-0052-5_3, © Springer Basel AG 2010

**Definition 3.1.2.** A dendrex $x \in X_T$ is called *degenerate* if there exists a dendrex $y \in X_S$ and a degeneracy $\sigma \colon T \longrightarrow S$ such that $\sigma^*(y) = x$.

There are canonical inclusions and evident restriction functors

which all have left and right adjoints

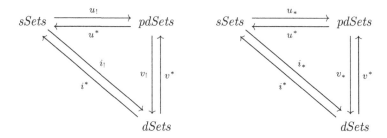

given by the corresponding Kan extensions. For example, the functor $i^*$ sends a dendroidal set $X$ to the simplicial set

$$i^*(X)_n = X_{i([n])}.$$

Its left adjoint $i_! \colon sSets \longrightarrow dSets$ is 'extension by zero', and sends a simplicial set $X$ to the dendroidal set given by

$$i_!(X)_T = \begin{cases} X_n & \text{if } T \cong i([n]), \\ \emptyset & \text{otherwise.} \end{cases}$$

It follows that $i_!$ is full and faithful and that $i^* i_!$ is the identity functor on simplicial sets.

*Example* 3.1.3. Let $T$ be a tree. The *standard $T$-dendrex* is the representable presheaf $\Omega(-,T)$. We will denote it by $\Omega[T]$ (just like $\Delta[n]$ in *sSets*). Explicitly, we have that

$$\Omega[T]_S = \Omega(S,T)$$

for every tree $S$. The relation $i_!(\Delta[n]) = \Omega[i([n])]$ holds for every $n$.

By the Yoneda Lemma, each dendrex $x$ of shape $T$ in a dendroidal set $X$ corresponds bijectively to a map of dendroidal sets $\hat{x} \colon \Omega[T] \longrightarrow X$. Note that $\Omega[-]$ is functorial, i.e., if $\alpha \colon S \longrightarrow T$ is a map of dendroidal sets then we have an induced map $\Omega[\alpha] \colon \Omega[S] \longrightarrow \Omega[T]$.

*Example* 3.1.4. The functor $\Omega \longrightarrow Oper$ which sends a tree $T$ to the coloured operad $\Omega(T)$ induces an adjunction

$$\tau_d : dSets \rightleftarrows Oper : N_d. \tag{3.1}$$

The functor $N_d$ is called the *dendroidal nerve*. Explicitly, for any operad $P$ the dendroidal nerve of $P$ is the dendroidal set

$$N_d(P)_T = Oper(\Omega(T), P).$$

The dendroidal nerve functor is fully faithful and $N_d(\Omega(T)) = \Omega[T]$ for every $T$ in $\Omega$. It extends the usual nerve functor from categories to simplicial sets. If $\mathcal{E}$ is any monoidal category and $\underline{\mathcal{E}}$ is the associated coloured operad (see Example 1.3.5), then

$$i^*(N_d(\underline{\mathcal{E}})) = N(\mathcal{E}).$$

For a dendroidal set $X$, we refer to the left adjoint $\tau_d(X)$ as the *operad generated by $X$*. It can be explicitly described as follows. For any dendroidal set $X$, the set of colours $col(\tau_d(X))$ is equal to $X_\eta$. The operations of the operad are generated by the elements of $X_{C_n}$, where $C_n$ is the $n$-th corolla, with the following relations:

(i)  $s(x_a) = \mathrm{id}_{x_a} \in \tau_d(X)(x_a; x_a)$ if $x_a \in X_\eta$ and $s = \sigma^*$, where $\sigma$ is the degeneracy $\sigma : C_1 \longrightarrow \eta$.

(ii) If $T$ is a tree of the form

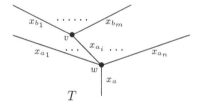

$$T$$

and $x \in X_T$, then $d_w(x) \circ_{x_{a_i}} d_v(x) = d_{x_{a_i}}(x)$, where

$$d_w(x) \in \tau_d(X)(x_{a_1}, \dots, x_{a_n}; x_a),$$
$$d_v(x) \in \tau_d(X)(x_{b_1}, \dots, x_{b_m}; x_{a_i}),$$
$$d_{x_{a_i}}(x) \in \tau_d(X)(x_{a_1}, \dots, x_{a_{i-1}}, x_{b_1}, \dots, x_{b_m}, x_{a_{i+1}}, \dots, x_{a_n}; x_a),$$

and $d_w = \partial_w^*$ is induced by the face map associated to removing the root vertex; $d_v = \partial_v^*$ is induced by the outer face map by cutting the upper part of the tree; and $d_{x_{a_i}} = \partial_{x_{a_i}}^*$ is induced by the inner face map by contracting the edge labeled $x_{a_i}$.

For example, $\tau_d(\Omega[T]) = \Omega(T)$ for every tree $T$ in $\Omega$.

The functor $\tau_d$ also extends the functor $\dot{\tau} \colon sSets \longrightarrow Cat$ left adjoint to the simplicial nerve, i.e., $\tau(X) = j^*\tau_d(i_!(X))$ for every simplicial set $X$. In particular, there is a diagram of adjoint functors

$$
\begin{array}{ccc}
sSets & \underset{i^*}{\overset{i_!}{\rightleftarrows}} & dSets \\
\tau \Big\Vert\, \big\uparrow N & & \tau_d \Big\Vert\, \big\uparrow N_d \\
Cat & \underset{j^*}{\overset{j_!}{\rightleftarrows}} & Oper
\end{array}
$$

with left adjoints on top or to the left. Moreover, the following commutativity relations hold up to natural isomorphisms:

$$\tau N = \mathrm{id}, \quad \tau_d N_d = \mathrm{id}, \quad i^* i_! = \mathrm{id}, \quad j^* j_! = \mathrm{id},$$

and

$$j_! \tau = \tau_d i_!, \quad N j^* = i^* N_d, \quad i_! N = N_d j_!.$$

There is also a column in the middle of the square relating planar dendroidal sets with non-$\Sigma$ operads.

*Remark* 3.1.5. Not everything commutes in the above diagram. The canonical map $\tau i^*(X) \longrightarrow j^* \tau_d(X)$ is not an isomorphism in general. This can be viewed, for example, by taking the representable dendroidal set $\Omega[T]$, where $T$ is the tree with three edges, one binary vertex and one nullary vertex:

Let $X = \partial_u \Omega[T] \cup \partial_v \Omega[T] \subseteq \partial \Omega[T]$ be the union of the outer faces. Then $i^*(X) = 0$. But $\tau_d(X) = \Omega(T)$, so $j^* \tau_d(X) \neq 0$.

Later we shall have to use that the Yoneda embedding $\Omega \longrightarrow dSets$, mapping a tree $T$ to the representable dendroidal set $\Omega[T]$, preserves pushouts of the form given in Lemma 2.3.3. We state this explicitly as follows.

**Proposition 3.1.6.** *Let the diagram*

$$
\begin{array}{ccc}
R & \overset{f}{\twoheadrightarrow} & S \\
g \downarrow & & \downarrow \\
T & \twoheadrightarrow & P
\end{array}
$$

*be a pushout square of surjections in* $\Omega$. *Then this pushout square is* absolute, *i.e.,*
*preserved by any functor. In particular, the induced square*

$$\Omega[R] \longrightarrow\!\!\!\!\rightarrow \Omega[S]$$
$$\downarrow \qquad\qquad \downarrow$$
$$\Omega[T] \longrightarrow\!\!\!\!\rightarrow \Omega[P]$$

*is a pushout square in* dSets.

Note that any surjection in $\Omega$ has a section, hence remains an epimorphism after applying the Yoneda embedding (or any other functor).

The proof is based on the well-known fact that *split coequalizers* are absolute [Mac71, Ch. VI, §6]. We recall that a diagram

$$A \underset{l}{\overset{k}{\rightrightarrows}} B \overset{p}{\longrightarrow} C$$

is *split* if there exist maps $t\colon C \longrightarrow B$ and $s\colon B \longrightarrow A$ such that $pt = \mathrm{id}_C$, $ks = \mathrm{id}_B$ and $tp = ls$. Any such split diagram is a coequalizer.

**Lemma 3.1.7.** *Consider a square*

$$
\begin{array}{ccc}
X & \overset{f}{\longrightarrow} & Y \\
{\scriptstyle g}\downarrow & & \downarrow{\scriptstyle u} \\
Z & \overset{v}{\longrightarrow} & P,
\end{array}
\qquad (3.2)
$$

*a section* $s\colon Z \longrightarrow X$ *of* $g$, *and the induced diagram*

$$X \underset{fsg}{\overset{f}{\rightrightarrows}} Y \overset{u}{\longrightarrow} P. \qquad (3.3)$$

(i) *The diagram* (3.2) *is a pushout if and only if* (3.3) *is a coequalizer.*

(ii) *If there are sections* $j\colon Y \longrightarrow X$ *of* $f$ *and* $t\colon P \longrightarrow Y$ *of* $u$ *satisfying the identity* $tu = (fsg)j$, *then* (3.3) *is a split coequalizer with 'splitting'*

$$X \overset{j}{\longleftarrow} Y \overset{t}{\longleftarrow} P.$$

*In particular, the pushout* (3.2) *is absolute if such sections* $s$, $t$ *and* $j$ *exist.*

*Proof.* Part (ii) is clear from the definition, and part (i) is an elementary diagram chase. By way of example, we prove that (3.2) is a pushout if (3.3) is a coequalizer. Take another object $W$, and arrows $\varphi\colon Y \longrightarrow W$ and $\psi\colon Z \longrightarrow W$ with $\varphi f = \psi g$. We look for a unique $\chi\colon P \longrightarrow W$ with $\chi u = \varphi$ and $\chi v = \psi$. Now $\varphi f sg = \psi gsg = \psi g = \varphi f$, so $\varphi$ factors uniquely through the coequalizer $u$ in (3.3) as $\varphi = \chi u$. Then also $\psi = \chi v$; indeed, $\chi v = \chi vgs = \chi ufs = \varphi fs = \psi gs = \psi$. $\qquad\square$

**Lemma 3.1.8.** *In any pushout square of degeneracies*

$$
\begin{array}{ccc}
R & \xrightarrow{\ \sigma_v\ } & S \\
\sigma_w \downarrow & & \downarrow \sigma_w \\
T & \xrightarrow{\ \sigma_v\ } & P
\end{array}
\tag{3.4}
$$

*in $\Omega$, there exist sections $s$, $t$ and $j$*

$$
\begin{array}{ccc}
R & \xleftarrow{\ j\ } & S \\
\uparrow s & & \uparrow t \\
T & & P
\end{array}
$$

*(of $\sigma_w, \sigma_w$ and $\sigma_v$ respectively) satisfying the equation*

$$
t\sigma_w = \sigma_v s\sigma_w j \colon S \longrightarrow S.
$$

*Proof.* Let $v$ and $w$ be the vertices

in $R$, with $v \neq w$. Sections of the maps in the pushout square (3.4) correspond to set-theoretic sections of sets of edges

$$
\begin{array}{ccc}
E(R) & \longrightarrow & E(S) \\
\downarrow & & \downarrow \\
E(T) & \longrightarrow & E(P).
\end{array}
$$

Since these sections are uniquely determined outside the edges $a$, $b$, $x$, $y$, it really comes down to finding sections in the following pushout diagram of sets:

$$
\begin{array}{ccc}
U & \longrightarrow & U/X \\
\downarrow & & \downarrow \\
U/A & \longrightarrow & V,
\end{array}
\tag{3.5}
$$

where $A = \{a, b\}$, $X = \{x, y\}$, and $U = A \cup X$. We can distinguish two cases:

If $A$ and $X$ are disjoint, then the diagram (3.5) looks like

$$
\begin{array}{ccc}
A + X & \xrightarrow{\ f\ } & A + \{1\} \\
\downarrow{\scriptstyle g} & & \downarrow{\scriptstyle u} \\
\{0\} + X & \xrightarrow{\ v\ } & \{0,1\}
\end{array}
$$

and we can take any sections $s$, $j$ and $t$,

$$
\begin{array}{ccc}
A + X & \xleftarrow{\ j\ } & A + \{1\} \\
\uparrow{\scriptstyle s} & & \uparrow{\scriptstyle t} \\
\{0\} + X & & \{0,1\},
\end{array}
$$

with $s(0) = t(0)$. Then $fsgj = tu$, as required.

If $A$ and $X$ are not disjoint, say $x = b$, then the diagram (3.5) looks like

$$
\begin{array}{ccc}
\{a, b = x, y\} & \xrightarrow{\ f\ } & \{a, b\} \\
\downarrow{\scriptstyle g} & & \downarrow{\scriptstyle u} \\
\{x, y\} & \xrightarrow{\ v\ } & \{0\}
\end{array}
$$

with $fa = a$, $fb = fy = b$ and $ga = gb$, $gy = y$. Then one can take sections $s$, $j$ and $t$,

$$
\begin{array}{ccc}
\{a, b, y\} & \xleftarrow{\ j\ } & \{a, b\} \\
\uparrow{\scriptstyle s} & & \uparrow{\scriptstyle t} \\
\{x, y\} & & \{0\},
\end{array}
$$

with $s(x) = b$, $t(0) = b$, $j(b) = b$, to obtain the identity $fsgj = tu$ again. $\qquad\square$

*Proof of Proposition* 3.1.6. It suffices (as in Lemma 2.3.3) to consider the case where the surjections $f$ and $g$ are degeneracies

$$
\sigma_v : R \longrightarrow S = R\backslash v \quad \text{and} \quad \sigma_w : R \longrightarrow T = R\backslash w.
$$

The proposition is evidently true in the case $v = w$. If $v \neq w$, then Lemma 3.1.8 completes the proof. $\qquad\square$

## 3.2 Faces, boundaries and horns

In this section we define face maps, boundaries and horns in the context of dendroidal sets.

**Definition 3.2.1.** Let $T$ be an object of $\Omega$ and $\alpha\colon S \longrightarrow T$ a face map in $\Omega$. The *$\alpha$-face* of $\Omega[T]$ is the dendroidal subset of $\Omega[T]$ given by the image of the map $\Omega[\alpha]\colon \Omega[S] \longrightarrow \Omega[T]$. We denote it by $\partial_\alpha\Omega[T]$. We write $\Phi_1(T)$ for the set of all faces of $T$.

Thus we have that

$$\partial_\alpha\Omega[T]_R = \{R \xrightarrow{\beta} S \xrightarrow{\alpha} T \text{ where } \beta \in \Omega[S]_R\}.$$

When the face map $\alpha$ is an inner face obtained by contracting an inner edge $e$, we denote $\partial_\alpha$ by $\partial_e$.

**Definition 3.2.2.** Let $T$ be an object of $\Omega$. The *boundary* of $\Omega[T]$ is the dendroidal subset $\partial\Omega[T]$ of $\Omega[T]$ obtained as the union of all possible faces of $\Omega[T]$. Namely,

$$\partial\Omega[T] = \bigcup_{\alpha\in\Phi_1(T)} \partial_\alpha\Omega[T].$$

If we take the union of all the faces except for one, we have the definition of a horn.

**Definition 3.2.3.** Let $T$ be an object of $\Omega$ and $\alpha \in \Phi_1(T)$ a face of $T$. The *$\alpha$-horn* of $\Omega[T]$ is the dendroidal subset $\Lambda^\alpha[T]$ of $\Omega[T]$ obtained as the union over all faces of $T$ except $\alpha$. That is,

$$\Lambda^\alpha[T] = \bigcup_{\beta\neq\alpha\in\Phi_1(T)} \partial_\beta\Omega[T].$$

As before, if $\alpha$ is an inner face map contracting an edge $e$, then we denote $\Lambda^\alpha[T]$ by $\Lambda^e[T]$. The horns of the form $\Lambda^e[T]$ are called *inner horns*. The other horns are called *outer horns*.

A *horn* in a dendroidal set $X$ is given by a map of dendroidal sets

$$\Lambda^\alpha[T] \longrightarrow X.$$

This horn is *inner* if $\Lambda^\alpha[T]$ is an inner horn and it is *outer* if $\Lambda^\alpha[T]$ is an outer horn.

The definitions of faces, boundaries and horns in dendroidal sets naturally extend the corresponding ones for simplicial sets. For example, if $\Lambda^k[n] \subseteq \Delta[n]$ denotes the simplicial $k$-horn, then the dendroidal set

$$i_!(\Lambda^k[n]) \subseteq i_!(\Delta[n]) = \Omega[L_n],$$

where $L_n$ denotes the linear tree with $n$ vertices and $n+1$ edges, is a horn in the dendroidal sense. Moreover, the horn $\Lambda^k[n]$ is an inner horn (i.e., $0 < k < n$) if and only if $i_!(\Lambda^k[n])$ is an inner horn.

Boundaries and horns can also be described in term of colimits. This extends, in the case of simplicial sets, the presentation of the boundary $\partial\Delta[n]$ and the horn $\Lambda^k[n]$ as a colimit of standard simplices.

Let $T_1 \longrightarrow T_2 \longrightarrow \cdots \longrightarrow T_n$ be a sequence of $n$ face maps in $\Omega$. The composition of these maps is called a *subface of codimension n* of $T_n$. Note that subfaces of codimension 1 are precisely the face maps. It follows from the dendroidal identities in Section 2.2.3 that every subface of a tree of codimension 2 decomposes in exactly two different ways as a composition of faces. Let $\Phi_2(T)$ be the set of all subfaces of codimension 2 of $T$. Thus, for each $\beta \colon S \longrightarrow T$ in $\Phi_2(T)$ there are exactly two maps $\beta_1 \colon S \longrightarrow T_1$ and $\beta_2 \colon S \longrightarrow T_2$ through which $\beta$ factors. Using $\beta_1$ and $\beta_2$, we can define two maps $\gamma_1$ and $\gamma_2$,

$$
\coprod_{(S \to T) \in \Phi_2(T)} \Omega[S] \xrightarrow[\gamma_2]{\gamma_1} \coprod_{(R \to T) \in \Phi_1(T)} \Omega[R],
$$

where the components of $\gamma_i$ are the compositions

$$
\Omega[S] \xrightarrow{\Omega[\beta_i]} \Omega[T_i] \longrightarrow \Omega[R]
$$

for each $\beta \colon S \longrightarrow T$ in $\Phi_2(T)$ and $i = 1, 2$.

**Lemma 3.2.4.** *Let $T$ be an object of $\Omega$. Then the boundary $\partial\Omega[T]$ can be obtained as the coequalizer*

$$
\coprod_{(S \to T) \in \Phi_2(T)} \Omega[S] \xrightarrow[\gamma_2]{\gamma_1} \coprod_{(R \to T) \in \Phi_1(T)} \Omega[R] \longrightarrow \partial\Omega[T].
$$

*Proof.* The universal property is verified by using the definition of $\partial\Omega[T]$ and the fact that every subface of codimension 2 decomposes exactly in two different ways as a composition of faces. $\square$

**Corollary 3.2.5.** *A map of dendroidal sets $\partial\Omega[T] \longrightarrow X$ corresponds exactly to a sequence of dendrices $\{x_R\}_{(R \to T) \in \Phi_1(T)}$ that agree on common faces, i.e., if $\beta \colon S \to T$ is a subface of codimension 2 which factors as*

$$
\begin{array}{ccc}
S & \xrightarrow{\beta_1} & T_1 \\
{\scriptstyle \beta_2}\big\downarrow & {\scriptstyle \beta}\searrow & \big\downarrow{\scriptstyle \alpha_1} \\
T_2 & \xrightarrow{\alpha_2} & T,
\end{array}
$$

*then $\beta_1^*(x_{T_1}) = \beta_2^*(x_{T_2})$.* $\square$

If $\alpha$ is a face of $T$, the $\alpha$-horn $\Lambda^\alpha[T]$ can be computed using the same coequalizer as before, but excluding the face $\alpha$.

**Lemma 3.2.6.** *Let $T$ be an object of $\Omega$ and $\alpha$ a face of $T$. Then the horn $\Lambda^\alpha[T]$ is the coequalizer*

$$\coprod_{(S\to T)\in\Phi_2(T)} \Omega[S] \underset{\gamma_2}{\overset{\gamma_1}{\rightrightarrows}} \coprod_{(R\to T)\neq\alpha\in\Phi_1(T)} \Omega[R] \longrightarrow \Lambda^\alpha[T].$$

*Proof.* The proof is analogous to that of Lemma 3.2.4. □

**Corollary 3.2.7.** *Let $\alpha$ be a face map in $T$. A horn $\Lambda^\alpha[T] \longrightarrow X$ in $X$ corresponds exactly to a sequence of dendrices $\{x_R\}_{(R\to T)\neq\alpha\in\Phi_1(T)}$ that agree on common faces, i.e., if $\beta\colon S \longrightarrow T$ is a subface of codimension 2 which factors as*

$$\begin{array}{ccc} S & \overset{\beta_1}{\longrightarrow} & T_1 \\ \beta_2\downarrow & \overset{\beta}{\searrow} & \downarrow\alpha_1 \\ T_2 & \underset{\alpha_2}{\longrightarrow} & T, \end{array}$$

*where $\alpha_1, \alpha_2 \neq \alpha$, then $\beta_1^*(x_{T_1}) = \beta_2^*(x_{T_2})$.* □

Finally, we will use the following terminology for dendrices in a dendroidal set. Let $\alpha\colon S \longrightarrow T$ be a map in $\Omega$, let $X$ be a dendroidal set, and let $t \in X_T$ be a $T$-dendrex. Consider the $S$-dendrex given by $\alpha^*(t)$. Then:

(i) $\alpha^*(t)$ is a *face* (resp. *inner face*, *outer face*) of $t$ if $\alpha$ is a face (resp. inner face, outer face) of $T$.

(ii) $\alpha^*(t)$ is a *subface* of $t$ if $\alpha$ is a subface of $T$.

(iii) $\alpha^*(t)$ is *isomorphic* to $t$ if $\alpha$ is an isomorphism.

(iv) $\alpha^*(t)$ is a *degeneracy* of $t$ is $\alpha$ is a composition of degeneracies.

## 3.3　Skeleta and coskeleta

Let $\Omega^{\leq n}$ denote the full subcategory of $\Omega$ consisting of trees with $n$ or less vertices. Similarly, one can define the full subcategory $\Delta^{\leq n}$ as the full subcategory of $\Delta$ with objects $[k]$ where $0 \leq k \leq n$. There is a commutative diagram

$$\begin{array}{ccc} \Delta^{\leq n} & \overset{j}{\longrightarrow} & \Omega^{\leq n} \\ i_n\downarrow & & \downarrow i_n \\ \Delta & \underset{i}{\longrightarrow} & \Omega, \end{array}$$

where $i_n$ denotes the fully faithful inclusion functor. The functors $i_n$ induce functors $i_n^*$ between the corresponding categories of presheaves and thus we have a commutative diagram

$$sSets^{\leq n} \xleftarrow{\ j^* \ } dSets^{\leq n} \qquad (3.6)$$

$$i_n^* \uparrow \qquad \qquad \uparrow i_n^*$$

$$sSets \xleftarrow{\ i^* \ } dSets$$

consisting of the inverse image functor of a pullback of presheaf toposes, together with the corresponding left adjoints $i_{n!}$, $j_!$ and $i_!$, and right adjoints $i_{n*}$, $j_*$ and $i_*$. Moreover, all $\alpha_*$ and $\alpha_!$ are full and faithful ($\alpha = i_n, j, i$).

**Definition 3.3.1.** Let $X$ be a dendroidal set. The *n-th skeleton* of $X$ is defined as $\mathrm{Sk}_n(X) = i_{n!}i_n^*(X)$ and the *n-th coskeleton* of $X$ as $\mathrm{coSk}_n(X) = i_{n*}i_n^*(X)$. (One can define with a similar formula the $n$-th skeleton of a simplicial set.)

There are natural morphisms

$$\mathrm{Sk}_n(X) \longrightarrow X \longrightarrow \mathrm{coSk}_n(X)$$

given by the counit and the unit of the corresponding adjunctions, for every dendroidal set $X$. There are also inclusions $\mathrm{Sk}_n(X) \subseteq \mathrm{Sk}_{n+1}(X)$ for every $n \geq 0$. It follows that $X = \cup_{n=0}^\infty \mathrm{Sk}_n(X)$, and this presentation of $X$ is called the *skeletal filtration* of $X$. One defines similarly the *coskeletal filtration* of a dendroidal set. Recall that the functor $i_n^* \colon dSets \longrightarrow dSets^{\leq n}$ is defined on representables as

$$i_n^*(\Omega[T]) = \Omega(i_n(-), T)$$

for every $T$ in $\Omega^{\leq n}$. Its left adjoint $i_{n!}$ is defined on representables as

$$i_{n!}(\Omega^{\leq n}[T]) = \Omega(-, i_n(T))$$

for every $T$ in $\Omega$. The fact that $i_n$ is fully faithful implies that $i_n^* i_{n!}(\Omega[T]) = \Omega[T]$ for every object $T$ in $\Omega$. Hence, $i_n^* i_{n!}(X) = X$ for all $X$, since every dendroidal set is a canonical colimit of representables. Using the adjunction one can check that also $i_n^* i_{n*}(X) = X$ for all $X$. It follows that $i_{n!}$ and $i_{n*}$ are both fully faithful.

**Definition 3.3.2.** A dendroidal set $X$ is called *n-coskeletal* if $X = \mathrm{coSk}_n(X)$.

**Proposition 3.3.3.** *A dendroidal set $X$ is n-coskeletal if, for every dendroidal set $Y$, each map $\mathrm{Sk}_n(Y) \longrightarrow X$ extends uniquely along $\mathrm{Sk}_n(Y) \longrightarrow Y$ to a map $Y \longrightarrow X$.*

*Proof.* Since $X$ is $n$-coskeletal, $X = \mathrm{coSk}_n(X)$. By an adjointness argument, there is a bijection between the sets of maps $dSets(Y, X)$ and $dSets(\mathrm{Sk}_n(Y), X)$. $\square$

If we make the definition of the Kan extension $i_{n!}$ explicit, we find that, for any tree $R$ in $\Omega$,

$$\mathrm{Sk}_n(X)_R = \varinjlim_{(T,\alpha)} \Omega(R, T),$$

where the colimit ranges over all trees $T$ with at most $n$ vertices and all $\alpha \in X_T$, i.e., maps $\alpha\colon \Omega[T] \longrightarrow X$. In other words, $\mathrm{Sk}_n(X)_R$ consists of equivalence classes of pairs $(\alpha, u)$, with $u\colon R \longrightarrow T$ in $\Omega$ and $\alpha\colon \Omega[T] \longrightarrow X$ in $dSets$, and $|V(T)| \le n$. The equivalence relation on such pairs is generated by

$$(\alpha v, u) \sim (\alpha, vu)$$

where $R \xrightarrow{u} T' \xrightarrow{v} T$ and $\Omega[T] \xrightarrow{\alpha} X$. The counit maps the equivalence class of $(\alpha, u)$ to $u^*(\alpha)$, or, in another notation, it maps $(\alpha, u)$ to the composition $\alpha \circ u$,

$$\Omega[R] \xrightarrow{u} \Omega[T] \xrightarrow{\alpha} X.$$

**Lemma 3.3.4.** *For each $n \ge 0$, the counit of the adjunction $\mathrm{Sk}_n(X) \longrightarrow X$ is a monomorphism for every dendroidal set $X$.*

*Proof.* To show that $\mathrm{Sk}_n(X) \longrightarrow X$ is injective, we need to prove that, if a diagram of the form

$$
\begin{array}{ccc}
\Omega[S] & \xrightarrow{\ \alpha\ } & X \\
\uparrow{\scriptstyle u} & & \uparrow{\scriptstyle \beta} \\
\Omega[R] & \xrightarrow{\ v\ } & \Omega[T]
\end{array}
$$

in $dSets$ commutes, where $S$ and $T$ have at most $n$ vertices, then $(\alpha, u) \sim (\beta, v)$. To see this, factor $u = if$ and $v = jg$ as in Lemma 2.3.2, and take the pushout $P$ in $\Omega$,

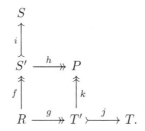

Then the functor $\Omega[-]$ preserves this pushout by Proposition 3.1.6, so we obtain a diagram in $dSets$ of the form

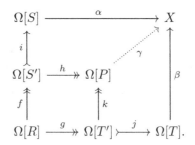

Now $S'$, $P$ and $T'$ have at most $n$ vertices since $S$ and $T$ do, so we can use the equivalence relation defining $\mathrm{Sk}_n(X)$, and find

$$(\alpha, u) = (\alpha, if) \sim (\alpha i, f) = (\gamma h, f) \sim (\gamma, hf),$$

and, in exactly the same way, $(\beta, v) \sim (\gamma, kg)$. Since $hf = kg$, this shows that $(\alpha, u) \sim (\beta, v)$ and proves the lemma. $\qquad\square$

The following proposition relates the skeleton and coskeleton constructions between the category of simplicial sets and the category of dendroidal sets.

**Proposition 3.3.5.** *The following relations hold:*

(i) $i^*(\mathrm{Sk}_n(X)) = \mathrm{Sk}_n(i^*(X))$ *and* $i^*(\mathrm{coSk}_n(X)) = \mathrm{coSk}_n(i^*(X))$ *for every dendroidal set $X$ and every $n \geq 0$.*

(ii) $i_!(\mathrm{Sk}_n(X)) = \mathrm{Sk}_n(i_!(X))$ *and* $i_*(\mathrm{coSk}_n(X)) = \mathrm{coSk}_n(i_*(X))$ *for every simplicial set $X$ and every $n \geq 0$.*

*Proof.* The proof is straightforward by using the following commutativity relations between the functors involved:

$$i^* i_{n!} = i_{n!} j^*, \quad j_* i_n^* = i_n^* i_*, \quad i_n^* i_! = j_! i_n^*, \quad i^* i_{n*} = i_{n*} j^*,$$

which follow from the fact that (3.6) is a pullback of presheaf toposes. $\qquad\square$

## 3.4 Normal monomorphisms

Recall from Definition 3.1.2 that a dendrex $t \in X_T$ is called degenerate if $t = \sigma(s)$ where $\sigma$ is a composition of degeneracies and $s$ is another dendrex. Any dendrex $t \in X_T$ where $T$ is a tree with no unary vertices is non-degenerate.

**Lemma 3.4.1.** *Any dendrex $x \in X_T$ is the restriction $\sigma^*(y)$ of a non-degenerate dendrex $y \in X_R$ along a surjection $\sigma \colon T \twoheadrightarrow R$ in $\Omega$. Moreover, given $x$, the map $\sigma$ and the dendrex $y$ are unique up to isomorphism.*

*Proof.* By the Yoneda Lemma, $x$ corresponds to a map $\hat{x} \colon \Omega[T] \longrightarrow X$. Consider, among all possible factorizations

$$\Omega[T] \xrightarrow{\ \sigma\ } \Omega[R] \xrightarrow{\ \hat{y}\ } X$$

of $\hat{x}$, those where $R$ has a minimal number of vertices, so that $y$ is necessarily non-degenerate. It suffices to show that any two such 'minimal' factorizations are isomorphic. But, given another one,

$$\Omega[T] \xrightarrow{\ \sigma'\ } \Omega[R] \xrightarrow{\ \hat{y}'\ } X,$$

we can form the pushout

$$
\begin{array}{ccc}
T & \overset{\sigma'}{\twoheadrightarrow} & R' \\
\sigma \downarrow & & \downarrow \tau' \\
R & \underset{\tau}{\twoheadrightarrow} & P
\end{array}
$$

in $\Omega$ by Lemma 2.3.3. Since $\Omega[-]$ of this pushout is a pushout in dendroidal sets (by Proposition 3.1.6) and since $\hat{y}\sigma = \hat{x} = \hat{y}'\sigma'$ by assumption, we find $\hat{z} \colon \Omega[P] \longrightarrow X$ with $\hat{z}\tau = \hat{y}$ and $\hat{z}\tau' = \hat{y}'$. But then, by minimality of $R$ and $R'$, both $\tau$ and $\tau'$ must be isomorphisms. Thus, we have the following diagram:

$$
\begin{array}{ccc}
\Omega[T] & \overset{\sigma'}{\longrightarrow} & \Omega[R'] \\
\sigma \downarrow & \nearrow{\scriptstyle\sim} & \downarrow \hat{y}' \\
\Omega[R] & \underset{\hat{y}}{\longrightarrow} & X,
\end{array}
$$

and $(\sigma, \hat{y})$ and $(\sigma', \hat{y}')$ are isomorphic.                                          $\square$

For $n > 0$, we can consider the following commutative diagram:

$$
\begin{array}{ccc}
\coprod_{(T,t)} \partial\Omega[T] & \longrightarrow & \mathrm{Sk}_{n-1}(X) \\
\downarrow & & \downarrow \\
\coprod_{(T,t)} \Omega[T] & \longrightarrow & \mathrm{Sk}_n(X),
\end{array}
\tag{3.7}
$$

where the coproducts are taken over all isomorphism classes of pairs $(T,t)$ in the category of elements of $X$ such that $T$ has $n$ vertices and $t \in X_T$ is non-degenerate. For the case $n = 0$, note that $\mathrm{Sk}_0(X) = \coprod_{x \in X_\eta} \Omega[\eta]$.

**Definition 3.4.2.** A monomorphism $X \rightarrowtail Y$ in $dSets$ is called *normal* if, for every tree $T$ in $\Omega$, every non-degenerate element $y \in Y_T$ which is not in the image of $X_T$ has a trivial stabilizer $\mathrm{Aut}(T)_y \subseteq \mathrm{Aut}(T)$, where $\mathrm{Aut}(-)$ denotes the automorphism group of the corresponding tree. An object $X$ is called *normal* if the map $\emptyset \rightarrowtail X$ is a normal monomorphism.

*Example* 3.4.3. For every tree $T$ in $\Omega$, the representable dendroidal set $\Omega[T]$ is normal. If $\sigma \colon X \rightarrowtail Y$ is a monomorphism and $Y$ is normal, then $\sigma$ is a normal monomorphism.

The skeletal filtration of a dendroidal set is called *normal* if the diagram (3.7) is a pushout. This property gives a characterization of normal dendroidal sets:

**Proposition 3.4.4.** *A dendroidal set is normal if and only if it admits a normal skeletal filtration.*

*Proof.* If a dendroidal set $X$ admits a normal skeletal filtration, it is easy to see that $X$ is normal, and the proof is left to the reader.

We prove in detail the other direction. Let us fix a representing element $(T, \beta)$ in each isomorphism class $[(T, \beta)]$ of non-degenerate dendrices $\beta \in X_T$, i.e., $\hat{\beta} \colon \Omega[T] \longrightarrow X$. We begin by observing that, in the category of sets, a pullback diagram of the form

$$
\begin{array}{ccc}
A & \xrightarrow{\ q\ } & C \\
\scriptstyle n\ \downarrow & & \downarrow\ \scriptstyle m \\
B & \xrightarrow[\ p\ ]{} & D,
\end{array}
$$

with $p$ an epimorphism and $m$ a monomorphism as indicated, is a pushout if and only if, as subsets of $B \times B$, the set $B \times_D B$ is contained in the union of the diagonal $B \to B \times B$ and $A \times_C A \rightarrowtail B \times_D B$. Since pullbacks and pushouts in presheaf categories are computed pointwise, the same observation applies to pushout diagrams in the category of dendroidal sets. Let us now check that the diagram

$$
\begin{array}{ccc}
\partial P = \coprod \partial\Omega[T] & \longrightarrow & \mathrm{Sk}_{n-1}(X) \\
\downarrow & & \downarrow \\
P = \coprod \Omega[T] & \longrightarrow & \mathrm{Sk}_n(X) \rightarrowtail X
\end{array}
$$

is a pushout. The square is clearly a pullback and we know that $\mathrm{Sk}_n(X) \longrightarrow X$ is a monomorphism for each $n$, by Lemma 3.3.4. Hence it is enough to prove that $P \times_X P \rightarrowtail P \times P$ is contained in the union of the diagonal $P \longrightarrow P \times P$ and $\partial P \times_X \partial P \rightarrowtail P \times P$. To this end, fix one representative $(T, \alpha)$ in each isomorphism class $[(T, \alpha)]$ (these classes index the coproduct in the diagram). For a tree $R$, an element $\xi \in (P \times_X P)_R$ is a commutative square

$$
\begin{array}{ccc}
\Omega[S] & \xrightarrow{\ \alpha\ } & X \\
\scriptstyle u\ \uparrow & & \uparrow\ \scriptstyle \beta \\
\Omega[R] & \xrightarrow[\ v\ ]{} & \Omega[T],
\end{array}
$$

where $(S, \alpha)$ and $(T, \beta)$ are representatives as above; in particular, $\alpha \in X_S$ and $\beta \in X_T$ are non-degenerate. If neither $u \colon R \longrightarrow S$ nor $v \colon R \longrightarrow T$ is surjective, then $u$ and $v$ factor through $\partial\Omega[S]$ and $\partial\Omega[T]$, so the element $\xi \in (P \times_X P)_R$ in fact lies in $(\partial P \times_X \partial P)_R$. Hence we may assume that one of $u$ and $v$ is surjective; say $u$ is. Now factor $v = gj$ as an epimorphism followed by a monomorphism, and form the pushout

$$\begin{array}{ccc} S & \xrightarrow{\;h\;} & P \\ {\scriptstyle u}\big\uparrow & & \big\uparrow{\scriptstyle k} \\ R & \xrightarrow[\;g\;]{} & T' \xrightarrow{\;j\;} T. \end{array}$$

We use Proposition 3.1.6 to find a diagram

$$\begin{array}{ccccc} \Omega[S] & \xrightarrow{\;h\;} & \Omega[P] & \xrightarrow{\;\gamma\;} & X \\ {\scriptstyle u}\big\uparrow & & \big\uparrow{\scriptstyle k} & & \big\uparrow{\scriptstyle \beta} \\ \Omega[R] & \xrightarrow[\;g\;]{} & \Omega[T'] & \xrightarrow{\;j\;} & \Omega[T] \end{array}$$

with $\gamma h = \alpha$. But $\alpha$ is non-degenerate, so $h$ must be an isomorphism. But then we have maps

$$T \xleftarrow{\;j\;} T' \xrightarrow{\;h^{-1}k\;} S,$$

where $S$ and $T$ have exactly $n$ vertices. So $j$ must be an isomorphism, as must be $k$. Thus, writing $\theta = h^{-1}k$, we have

$$\begin{array}{ccc} \Omega[S] & \xrightarrow{\;\alpha\;} & X \\ {\scriptstyle u}\big\uparrow {\scriptstyle \theta}\!\!\diagdown & & \big\uparrow{\scriptstyle \beta} \\ \Omega[R] & \xrightarrow[\;v\;]{} & \Omega[T]. \end{array}$$

In other words, $(S, \alpha)$ and $(T, \beta)$ lie in the same isomorphism class, hence they must be equal by our choice of representatives. But then $\theta = \mathrm{id}$, since $X$ is assumed normal. We conclude that $u = v$ as well, so the element $\xi \in P \times_X P$ represented by $(\alpha, u)$ and $(\beta, v)$ lies in fact in the diagonal. This completes the proof.   □

**Lemma 3.4.5.** *Let $p \colon Y \longrightarrow X$ be a map of dendroidal sets, and assume that $X$ is normal. Then $Y$ is normal as well.*

*Proof.* First recall that any normal object has a nice skeletal filtration, i.e., can be built up by attaching cells $x$ by pushouts of the form

$$\begin{array}{ccc} \partial\Omega[T] & \longrightarrow & A \\ \big\downarrow & & \big\downarrow \\ \Omega[T] & \xrightarrow{\;x\;} & B. \end{array}$$

Such an attached cell $x$ always comes with a map $\Omega[T] \xrightarrow{\;x\;} B$ which is 'injective on its interior' (just like for usual cell complexes). Indeed, by looking at pushouts of this form in *Sets* as before, one sees that the kernel pair $\Omega[T] \times_B \Omega[T] \subseteq \Omega[T] \times \Omega[T]$ of the map $x$ is the union of the diagonal $\Omega[T]$ and $\partial\Omega[T] \times_A \partial\Omega[T]$. This simple observation implies that, for any non-degenerate dendrex $x \in X(T)$ in a normal dendroidal set $X$, the corresponding map $\Omega[T] \xrightarrow{x} X$ has the cancellation property with respect to epimorphisms in $\Omega$:

$$\text{if } \Omega[R] \underset{\gamma}{\overset{\beta}{\rightrightarrows}} \Omega[T] \xrightarrow{x} X \text{ and } x\beta = x\gamma, \text{ then } \beta = \gamma.$$

To prove the lemma, let $y \in Y(T)$ be non-degenerate and suppose that $\alpha \in \mathrm{Aut}(T)$ fixes $y$, with $\alpha \neq 1$. Then $\alpha$ also fixes $py$, so $py$ must be degenerate by the assumption that $X$ is normal; say that $py = \rho^* x$, where $\rho\colon T \twoheadrightarrow R$ and $x \in X(R)$ is non-degenerate,

$$
\begin{array}{ccc}
T \xrightarrow{\ \underset{\sim}{\alpha}\ } T \xrightarrow{\ y\ } Y \\
\ \ \downarrow{\scriptstyle \rho} \qquad\quad \downarrow{\scriptstyle p} \\
R \xrightarrow{\ \ x\ \ } X.
\end{array}
$$

Then $x(\rho\alpha) = x\rho$ because $y\alpha = y$; hence, $\rho\alpha = \rho$ by the cancellation property. This means that $\alpha$ is an automorphism of $T$ which permutes the edges on each of the fibers of $\rho$. But these fibers are linear, so $\alpha$ must be the identity. □

# Lecture 4

# Tensor product of dendroidal sets

Like any category of presheaves, the category of dendroidal sets is cartesian closed. In this lecture, we will discuss another monoidal structure, which is also closed, and seems more relevant than the cartesian structure. It is closely related to the tensor product of operads introduced by Boardman and Vogt, and it makes the embedding of simplicial sets into dendroidal sets into a strong monoidal functor.

## 4.1 The Boardman–Vogt tensor product

The category of small categories *Cat* is a cartesian closed category for which the internal hom $\mathrm{Hom}_{Cat}(\mathcal{C}, \mathcal{D})$ between two categories $\mathcal{C}$ and $\mathcal{D}$ is defined as the category whose objects are functors from $\mathcal{C}$ to $\mathcal{D}$ and whose morphisms are natural transformations between them. In this section, we show that the category of coloured operads is a closed symmetric monoidal category with the so-called Boardman–Vogt tensor product [BV73, Definition 2.14].

We recall the definition of the Boardman–Vogt tensor product for coloured operads.

**Definition 4.1.1.** Let $P$ be a symmetric $C$-coloured operad and let $Q$ be a symmetric $D$-coloured operad. The *Boardman–Vogt tensor product* $P \otimes_{BV} Q$ is a $(C \times D)$-coloured operad defined in terms of generators and relations in the following way. For each $d \in D$ and each operation $p \in P(c_1, \ldots, c_n; c)$ there is a generator

$$p \otimes d \in P \otimes_{BV} Q((c_1, d), \ldots, (c_n, d); (c, d)).$$

Similarly, for each $c \in C$ and each $q \in Q(d_1, \ldots, d_m; d)$ there is a generator

$$c \otimes q \in P \otimes_{BV} Q((c, d_1), \ldots, (c, d_m); (c, d)).$$

I. Moerdijk and B. Toën, *Simplicial Methods for Operads and Algebraic Geometry*, Advanced Courses in Mathematics - CRM Barcelona, DOI 10.1007/978-3-0348-0052-5_4, © Springer Basel AG 2010

These generators are subject to the following relations:

(i)  $(p \otimes d) \circ ((p_1 \otimes d), \ldots, (p_n \otimes d)) = (p \circ (p_1, \ldots, p_n)) \otimes d.$

(ii)  $\sigma^*(p \otimes d) = (\sigma^* p) \otimes d$ for every $\sigma \in \Sigma_n$.

(iii)  $(c \otimes q) \circ ((c \otimes q_1), \ldots, (c \otimes q_m)) = c \otimes (q \circ (q_1, \ldots, q_m)).$

(iv)  $\sigma^*(c \otimes q) = c \otimes (\sigma^* q)$ for every $\sigma \in \Sigma_m$.

(v)  $\sigma_{n,m}^* ((p \otimes d) \circ ((c_1 \otimes q), \ldots, (c_n \otimes q))) = (c \otimes q) \circ ((p \otimes d_1), \ldots, (p \otimes d_m)),$
where $\sigma_{n,m} \in \Sigma_{nm}$ is the permutation described as follows. Consider $\Sigma_{nm}$ as the set of bijections of the set $\{0, 1, \ldots, nm - 1\}$. Each element of this set can be written uniquely in the form $kn + j$ where $0 \le k < m$ and $0 \le j < n$ as well as in the form $km + j$ where $0 \le k < n$ and $0 \le j < m$. The permutation $\sigma_{n,m}$ is then defined by $\sigma_{n,m}(kn + j) = jm + k$.

Observe that relations (i) and (ii) imply that for every $d \in D$ the map $P \longrightarrow P \otimes_{BV} Q$ given by $p \longmapsto p \otimes d$ is a map of operads. Similarly, relations (iii) and (iv) ensure that for every $c \in C$ the map $Q \longrightarrow P \otimes_{BV} Q$ given by $q \longmapsto c \otimes q$ is a map of operads.

*Example* 4.1.2. We illustrate relation (v), called the *interchange relation*, with the following examples. Suppose that $n = 2$ and $m = 3$. The left-hand operation of relation (v), before applying $\sigma_{2,3}^*$, can be represented by the tree

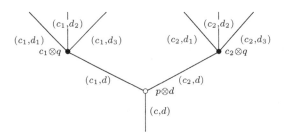

The right-hand operation can be represented by the tree

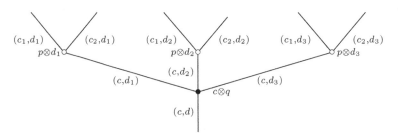

And the permutation $\sigma_{2,3}$ corresponds to the permutation $(2\ 4\ 5\ 3)$ of $\Sigma_6$. We represent the vertices coming from operations in $P$ by a white dot $\circ$ and the vertices coming from operations in $Q$ by a black dot $\bullet$.

The Boardman–Vogt tensor product preserves colimits in each variable separately. In fact, there is a corresponding internal hom making the category of coloured operads closed monoidal.

**Theorem 4.1.3.** *The category Oper with the Boardman–Vogt tensor product $\otimes_{BV}$ is a closed symmetric monoidal category.*

*Proof.* The unit for the tensor product is the initial operad $I$ on one colour, i.e., $I(*, *) = \{*\}$ and the empty set otherwise. It follows from the definition that this tensor product is associative, commutative and unital. This proves that *Oper* is symmetric monoidal.

We define the internal hom for coloured operads as follows. Let $P$ be a $C$-coloured operad and $Q$ a $D$-coloured operad. Then $\mathrm{Hom}_{Oper}(P, Q)$ is the operad whose colours are the maps of operads $P \longrightarrow Q$, and if $\alpha_1, \ldots, \alpha_n, \beta$ are $n + 1$ such maps, then the elements of

$$\mathrm{Hom}_{Oper}(P, Q)(\alpha_1, \ldots, \alpha_n; \beta)$$

are maps $f$ assigning to each colour $c \in C$ an element $f_c \in Q(\alpha_1 c, \ldots, \alpha_n c; \beta c)$. The maps $f_c$ should be natural with respect to all the operations in $P$. For example, if $p \in P(c_1, c_2; c)$, then

$$\beta(p)(f_{c_1}, f_{c_2}) \in Q(\alpha_1 c_1, \ldots, \alpha_n c_1, \alpha_1 c_2, \ldots, \alpha_n c_2; \beta c)$$

is the image under a suitable permutation of

$$f_c(\alpha_1(p), \ldots, \alpha_n(p)) \in Q(\alpha_1 c_1, \alpha_1 c_2, \ldots, \alpha_n c_1, \alpha_n c_2; \beta c).$$

We need to construct a bijection

$$Oper(P \otimes_{BV} Q, R) \cong Oper(P, \mathrm{Hom}_{Oper}(Q, R))$$

natural in $P$, $Q$ and $R$.

Let $\varphi \colon P \otimes_{BV} Q \longrightarrow R$ be a map of coloured operads. For each $c \in C$ we have a map of operads $\varphi_c$ defined by the composition

$$Q \longrightarrow P \otimes_{BV} Q \xrightarrow{\varphi} R$$

where the first map sends $q$ to $c \otimes q$. This defines a map from the colours of $P$ to the colours of $\mathrm{Hom}_{Oper}(Q, R)$. Now, if we have an operation $p \in P(c_1, \ldots, c_n; c)$, we define

$$f_d = \varphi(p \otimes d) \in R(\varphi_{c_1} d, \ldots, \varphi_{c_n} d; \varphi_c d)$$

for every $d \in D$.

Conversely, let $\psi \colon P \longrightarrow \mathrm{Hom}_{Oper}(Q, R)$ be a map of coloured operads. To construct a map $\overline{\psi} \colon P \otimes_{BV} Q \longrightarrow R$, we need to define it on the colours and the generators of $P \otimes_{BV} Q$. If $(c, d) \in C \times D$, then $\overline{\psi}(c, d) = \psi(c)(d)$. For a generator of the form $c \otimes q$, where $q \in Q(d_1, \ldots, d_n; d)$, we define

$$\overline{\psi}(c \otimes q) = \psi(c)(q).$$

For a generator of the form $p \otimes d$, where $p \in P(c_1, \ldots, c_n; c)$, we define

$$\overline{\psi}(p \otimes d) = \psi(p)_d.$$

It is now easy to check that $\overline{\psi}$ thus defined is compatible with the relations of the Boardman–Vogt tensor product.                                                   $\square$

*Remark* 4.1.4. Note that in the definition of the Boardman–Vogt tensor product it is crucial that the coloured operads involved are symmetric. However, the tensor product still makes sense without the symmetries when one of the operads involved has only unary operations.

## 4.2   Tensor product of dendroidal sets

The category of dendroidal sets is a category of presheaves, hence cartesian closed. The cartesian product of dendroidal sets extends the cartesian product of simplicial sets, i.e.,

$$i_!(X \times Y) \cong i_!(X) \times i_!(Y)$$

for every two simplicial sets $X$ and $Y$. (Note, however, that $i_!$ does not preserve the terminal object.)

As mentioned before, there is another closed monoidal structure on *dSets*, strongly related with the Boardman–Vogt tensor product of coloured operads. For any two trees $T$ and $S$ in $\Omega$, the tensor product of the representables $\Omega[T]$ and $\Omega[S]$ is defined as

$$\Omega[T] \otimes \Omega[S] = N_d(\Omega(T) \otimes_{BV} \Omega(S)),$$

where $N_d$ is the dendroidal nerve functor (see Example 3.1.4), $\Omega(T)$ and $\Omega(S)$ are the coloured operads associated to the trees $T$ and $S$ respectively (see Section 2.3), and $\otimes_{BV}$ is the Boardman–Vogt tensor product.

This defines a tensor product in the whole category of dendroidal sets, since, being a category of presheaves, every object is a canonical colimit of representables and $\otimes$ preserves colimits in each variable.

**Definition 4.2.1.** Let $X$ and $Y$ be two dendroidal sets and let $X = \varinjlim \Omega[T]$ and $Y = \varinjlim \Omega[S]$ be their canonical expressions as colimits of representables. Then the *tensor product* $X \otimes Y$ is defined as

$$X \otimes Y = \varinjlim \Omega[T] \otimes \varinjlim \Omega[S] = \varinjlim N_d(\Omega(T) \otimes_{BV} \Omega(S)).$$

It follows from general category theory [Kel82] that this tensor product is automatically closed, and that the set of $T$-dendrices of the internal hom is defined by

$$\text{Hom}_{dSets}(X,Y)_T = dSets(\Omega[T] \otimes X, Y),$$

for every two dendroidal sets $X$ and $Y$ and every $T$ in $\Omega$. The dendroidal structure of $\text{Hom}_{dSets}(X,Y)$ is given in the obvious way.

**Theorem 4.2.2.** *The category of dendroidal sets admits a closed symmetric monoidal structure. This monoidal structure is uniquely determined (up to isomorphism) by the property that there is a natural isomorphism*

$$\Omega[T] \otimes \Omega[S] \cong N_d((\Omega(T) \otimes_{BV} \Omega(S))$$

*for any two objects $T$ and $S$ of $\Omega$. The unit of the tensor product is the representable dendroidal set $\Omega[\eta] = i_!(\Delta[0]) = U$.* $\qquad\square$

The following are some basic properties of the tensor product of dendroidal sets in relation to the Boardman–Vogt tensor product of coloured operads and to the cartesian product of simplicial sets.

**Proposition 4.2.3.** *The following properties hold:*

(i) *For any two simplicial sets $X$ and $Y$, there is a natural isomorphism*

$$i_!(X) \otimes i_!(Y) \cong i_!(X \times Y).$$

(ii) *For any two dendroidal sets $X$ and $Y$, there is a natural isomorphism*

$$\tau_d(X \otimes Y) \cong \tau_d(X) \otimes_{BV} \tau_d(Y).$$

(iii) *For any two coloured operads $P$ and $Q$, there is a natural isomorphism*

$$\tau_d(N_d(P) \otimes N_d(Q)) \cong P \otimes_{BV} Q.$$

*Proof.* To prove (i), it is enough to check that it holds for the representables in *sSets*. Note first that, if we view $[n]$ and $[m]$ in $\Delta$ as categories, then by using the linear order one has that

$$j_!([n] \times [m]) \cong j_!([n]) \otimes_{BV} j_!([m]).$$

Therefore, there is a chain of natural isomorphisms

$$i_!(\Delta[n] \times \Delta[m]) \cong i_!(N([n]) \times N([m])) \cong i_!(N([n] \times [m]))$$
$$\cong N_d j_!([n] \times [m]) \cong N_d(j_!([n]) \otimes_{BV} j_!([m])) \cong N_d(\Omega(L_n) \otimes_{BV} \Omega(L_m))$$
$$\cong \Omega[L_n] \otimes \Omega[L_m] \cong i_!(\Delta[n]) \otimes i_!(\Delta[m]),$$

where $L_n$ and $L_m$ denote the linear tree with $n$ and $m$ vertices and $n+1$ and $m+1$ edges respectively.

Again, to prove (ii) it suffices to do it for representables in *dSets*. But this is clear by using the natural isomorphism $\tau_d N_d \cong \mathrm{id}$. More precisely,

$$\tau_d(\Omega[T] \otimes \Omega[S]) \cong \tau_d N_d(\Omega(T) \otimes_{BV} \Omega(S))$$
$$\cong \Omega(T) \otimes_{BV} \Omega(S) \cong \tau_d(\Omega[T]) \otimes_{BV} \tau_d(\Omega[S]).$$

Part (iii) follows from part (ii) by using again that $\tau_d N_d \cong \mathrm{id}$ and replacing $X$ by $N_d(P)$ and $Y$ by $N_d(Q)$. □

*Remark* 4.2.4. There is no tensor product in the category of planar dendroidal sets coming from the Boardman–Vogt tensor product, since the latter is defined only for symmetric operads. However, as we have seen in Remark 4.1.4, the Boardman–Vogt tensor product still makes sense for non-symmetric operads when at least one of them has only unary operations. This means that, although we cannot define $X \otimes Y$ for planar dendroidal sets $X$ and $Y$ in general, we can define $u_!(K) \otimes Y$ where $K$ is any simplicial set and $Y$ is any planar dendroidal set. In fact, *pdSets* is a simplicial category with tensors and cotensors.

**Theorem 4.2.5.** *The category pdSets of planar dendroidal sets is enriched, tensored and cotensored over simplicial sets.*

*Proof.* Given two planar dendroidal sets $X$ and $Y$, the simplicial enrichment $\mathrm{Hom}(X,Y)$ is defined by

$$\mathrm{Hom}(X,Y)_n = pdSets(u_!(\Delta[n]) \otimes X, Y)$$

where $u_! : sSets \longrightarrow pdSets$ is the left adjoint to the functor $u^*$ induced by the inclusion $u : \Delta \longrightarrow \Omega_p$. If $K$ is any simplicial set, we define a tensor

$$K \otimes Y = u_!(K) \otimes Y$$

and a cotensor

$$(Y^K)_T = pdSets(\Omega_p[T] \otimes u_!(K), Y)$$

for every planar dendroidal set $Y$. □

Thus, the Boardman–Vogt tensor product makes *sSets* into a cartesian closed category, *pdSets* into a simplicial category with tensors and cotensors, and *dSets* into a closed symmetric monoidal category. In fact, if we consider the cartesian structures on *Cat* and *sSets*, the Boardman–Vogt tensor product on *Oper* and the tensor product of dendroidal sets, then in the commutative diagram

$$
\begin{array}{ccc}
sSets & \underset{i^*}{\overset{i_!}{\rightleftarrows}} & dSets \\
\tau \big\uparrow\big\downarrow N & & \tau_d \big\uparrow\big\downarrow N_d \\
Cat & \underset{j^*}{\overset{j_!}{\rightleftarrows}} & Oper
\end{array}
$$

the functors $i_!$, $N$, $\tau$, $j_!$ and $\tau_d$ are strong monoidal. However, $i^*$ and $j^*$ are not strong monoidal functors. For example, if we denote by $T_1$, $T_2$ and $T_3$ the following trees:

then $i^*(\Omega[T_1]) = \emptyset$ and $i^*(\Omega[T_2]) = \Delta[0]$, but $\Omega[T_1] \otimes \Omega[T_2] = \Omega[T_3]$ and

$$i^*(\Omega[T_3]) = \Delta[1] \cup_{\Delta[0]} \Delta[1].$$

*Remark* 4.2.6. If $\mathcal{E}$ is a complete and cocomplete monoidal category, then the category $\mathcal{E}^{\Omega^{\mathrm{op}}}$ of dendroidal objects in $\mathcal{E}$ also has a Boardman–Vogt type tensor product. For any two objects $X$ and $Y$ in $\mathcal{E}^{\Omega^{\mathrm{op}}}$, their tensor product is defined by the following formula:

$$(X \otimes Y)_T = \varinjlim_{\Omega[T] \to \Omega[R] \otimes \Omega[S]} X_R \otimes_{\mathcal{E}} Y_S,$$

where $\otimes_{\mathcal{E}}$ is the tensor product of $\mathcal{E}$. If $\mathcal{E}$ is closed, then so is $\mathcal{E}^{\Omega^{\mathrm{op}}}$ (see [MW07, Appendix A] and [BM08, §7]).

## 4.3 Shuffles of trees

In this section, we describe the tensor product $\Omega[S] \otimes \Omega[T]$ for any two trees in $\Omega$, in order to give a better understanding of the tensor product of dendroidal sets. Suppose first that $S = L_n$ and $T = L_m$ are linear trees. Then, by Proposition 4.2.3(i),

$$\Omega[L_n] \otimes \Omega[L_m] = i_!(\Delta[n]) \otimes i_!(\Delta[m]) \cong i_!(\Delta[n] \times \Delta[m]).$$

The non-degenerate simplices of a product of two representables in simplicial sets are computed by means of *shuffles*. An $(n, m)$-shuffle is a path of maximal length in the partially ordered set $[n] \times [m]$. The non-degenerate $(n + m)$-simplices of $\Delta[n] \times \Delta[m]$ correspond to $(n, m)$-shuffles. In fact,

$$\Delta[n] \times \Delta[m] = \bigcup_{(n,m)} \Delta[n + m],$$

where the union is taken over all possible $(n, m)$-shuffles.

*Example* 4.3.1. Let $n = 2$ and $m = 1$. There are three $(2, 1)$-shuffles in $[2] \times [1]$, namely $(00, 01, 02, 12)$, $(00, 01, 11, 12)$ and $(00, 10, 11, 12)$. If we picture $\Delta[2] \times \Delta[1]$

as the following prism

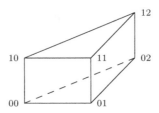

we can see that each $(2,1)$-shuffle corresponds to a tetrahedron, and that they give a decomposition of $\Delta[2] \times \Delta[1]$ as the union of three copies of $\Delta[3]$:

$$(00,01,02,12) \qquad (00,01,11,12) \qquad (00,10,11,12)$$

To give an explicit description of the tensor product of any two representables in $\Omega$, we need to introduce shuffles of trees. Recall that we denote by $E(T)$ the set of edges of a given tree $T$.

**Definition 4.3.2.** Let $S$ and $T$ be two objects of $\Omega$. A *shuffle* of $S$ and $T$ is a tree $R$ whose set of edges is a subset of $E(S) \times E(T)$. The root of $R$ is $(a, x)$ where $a$ is the root of $S$ and $x$ is the root of $T$, and its leaves are labelled by all pairs of the form $(l_S, l_T)$, where $l_S$ is a leaf of $S$ and $l_T$ is a leaf of $T$. Its vertices are either of the form

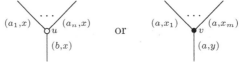

where $u$ is a vertex of $S$ with input edges $a_1, \dots, a_n$ and output $b$, and $v$ is a vertex of $T$ with input edges $x_1, \dots, x_m$ and output $y$. We will refer to the first type of vertices as *white vertices* and to the second type of vertices as *black vertices*. To make this distinction clear, we picture them as $\circ$ and $\bullet$ respectively.

Note that there is a bijection between the shuffles of two linear trees $L_n$ and $L_m$ and the $(n, m)$-shuffles of $[n] \times [m]$.

*Example* 4.3.3. Let $S$ and $T$ be the trees

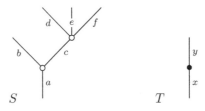

Then the set of shuffles of $S$ and $T$ consists of the following three trees:

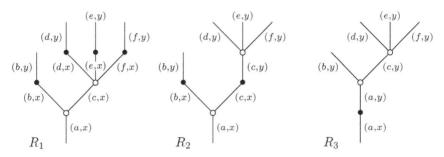

The set of shuffles of two trees $S$ and $T$ has a natural partial order. The minimal tree $R_1$ in this poset is the one obtained by stacking a copy of the black tree $T$ on top of each of the inputs of the white tree $S$. More precisely, on the bottom of $R_1$ there is a copy $S \otimes r_T$ of the tree $S$ all of whose edges are renamed $(a, r_T)$ where $r_T$ is the output edge at the root of $T$. For each input edge $b$ of $S$, a copy of $T$ is grafted on the edge $(b, r_T)$ of $S \otimes r_T$, with edges $x$ in $T$ renamed $(b, x)$. The maximal tree $R_N$ in the poset is the similar tree with copies of the white tree $S$ grafted on each of the input edges of the black tree. Schematically, the trees $R_1$ and $R_N$ look like

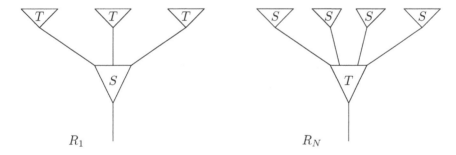

There are intermediate shuffles $R_k$ $(1 < k < N)$ between $R_1$ and $R_N$ obtained by letting the black vertices in $R_1$ slowly percolate in all possible ways towards the root of the tree. Shuffles are also called *percolation schemes*. The *percolation rule* or *percolation relation* can be made explicit as follows. Each $R_k$ is obtained from an earlier $R_l$ by replacing a configuration

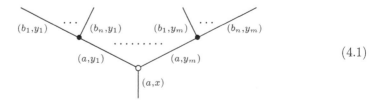

$$(4.1)$$

in $R_l$ by a configuration

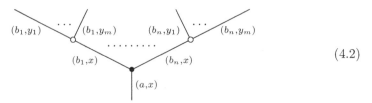

$$(4.2)$$

in $R_k$. If a shuffle $R_k$ is obtained from another $R_l$ by means of the above rule, then we say that $R_k$ is obtained by a *single percolation step* and denote this by $R_l \leq R_k$. This generates a partial order on the set of all shuffles.

It is important to make explicit the percolation relation above for trees with no input edges, i.e., $n = 0$ or $m = 0$. If $m = 0$ and $n \neq 0$, then we have the relation

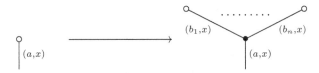

If $n = 0$ and $m \neq 0$, then we have the relation

And if $n = m = 0$, we have the relation

*Example* 4.3.4 **(Taken from [MW09, Example 9.4])**. Let $S$ and $T$ be the following two trees; here, we have singled out one particular edge $e$ in $S$, we have numbered the edges of $T$ as $1, \ldots, 5$, and denoted the colour $(e, i)$ in $R_j$ by $e_i$.

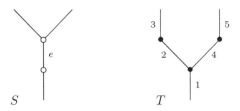

There are fourteen shuffles $R_1, \ldots, R_{14}$ of $S$ and $T$ in this case. Here is the

complete list of them:

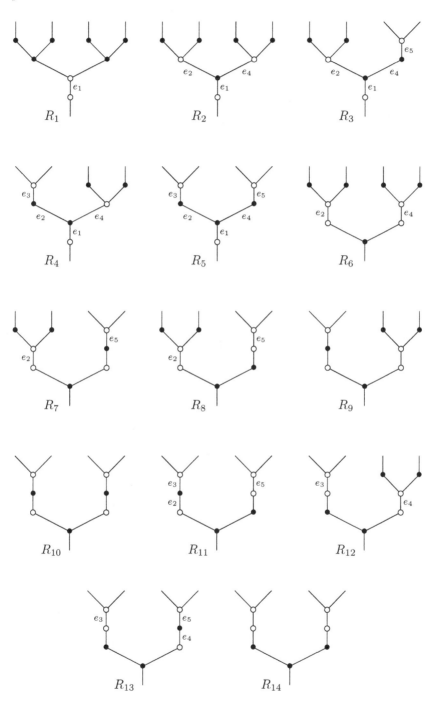

The poset structure on the shuffles above is:

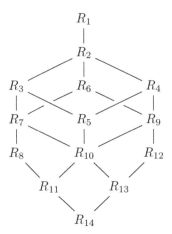

**Lemma 4.3.5.** *Every shuffle $R_i$ of $S$ and $T$ is equipped with a canonical monomorphism*

$$m \colon \Omega[R_i] \rightarrowtail \Omega[S] \otimes \Omega[T].$$

*The dendroidal subset given by the image of this monomorphism will be denoted by $m(R_i)$.*

*Proof.* The vertices of the dendroidal set $\Omega[R_i]$ are the edges of the tree $R_i$. The map $m$ is completely determined by asking it to map an edge named $(a, x)$ in $R_i$ to the vertex with the same name in $\Omega[S] \otimes \Omega[T]$. This map is a monomorphism. Indeed, any map

$$\Omega[R] \longrightarrow X$$

from a representable dendroidal set to an arbitrary one is a monomorphism as soon as the map $\Omega[R]_| \longrightarrow X_|$ on vertices is. $\square$

**Corollary 4.3.6.** *For any two objects $T$ and $S$ in $\Omega$, we have that*

$$\Omega[S] \otimes \Omega[T] = \bigcup_{i=1}^{N} m(R_i), \tag{4.3}$$

*where the union is taken over all possible shuffles of $S$ and $T$.*

The Boardman–Vogt relation says that if $R_k$ is obtained from $R_l$ by a single percolation step as above, then the image under $m$ of the face of $\Omega[R_k]$ obtained by contracting all the edges $(b_1, x), \ldots, (b_n, x)$ in (4.2) above coincides (as a subobject of $\Omega[S] \otimes \Omega[T]$) with the image of the face of $\Omega[R_l]$ obtained by contracting the edges $(a, y_1), \ldots, (a, y_m)$ in (4.1).

The following example illustrates that in the set of shuffles appearing in (4.3) some of them can be faces of others when one of the trees has vertices of valence zero. Thus, not all the shuffles are always needed in the union (4.3).

*Example* 4.3.7. Let $S$ and $T$ be the following two trees and observe that the tree $S$ has a vertex of valence zero.

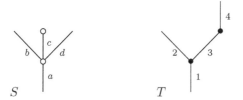

There are six shuffles $R_1, \ldots, R_6$ of $S$ and $T$. The colours $(e, k)$ of the edges in $R_i$ are denoted by $e_k$.

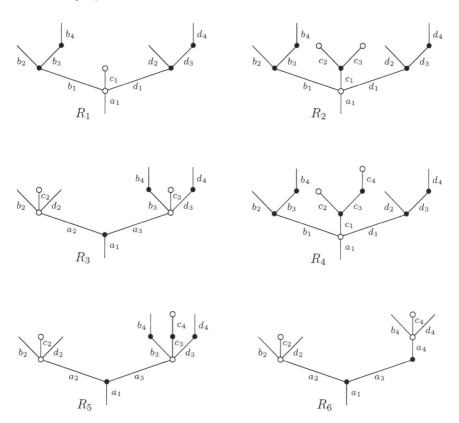

Observe that, in this case, $R_1$ is a face of $R_2$, which is a face of $R_4$. Similarly, $R_3$ is a face of $R_5$. Hence,

$$\Omega[S] \otimes \Omega[T] = m(R_4) \cup m(R_5) \cup m(R_6).$$

The poset structure on the shuffles above is the following:

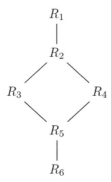

# Lecture 5

# A Reedy model structure on dendroidal spaces

In this lecture, which is based on [BM08], we extend the classical notion of Reedy category to include categories with non-trivial automorphisms, called *generalized Reedy categories*. We prove that, for any cofibrantly generated model category $\mathcal{E}$ and any generalized Reedy category $\mathbb{R}$, the category of functors $\mathcal{E}^{\mathbb{R}}$ carries a canonical model structure of Reedy type. Generalized Reedy categories include Segal's category $\Gamma$, Connes' cyclic category $\Lambda$, and the category of trees $\Omega$.

In particular, we describe a Reedy-like model structure on the category of dendroidal spaces $sSets^{\Omega^{\mathrm{op}}}$ in which the weak equivalences, fibrations and cofibrations can be described explicitly in terms of those in $sSets$.

## 5.1 Strict Reedy categories

In this section we recall the classical theory of Reedy categories and model structures on categories of diagrams over Reedy categories (see [Ree74], [Hir03, Ch. 15], [Hov99, §5.2]).

**Definition 5.1.1.** A *strict Reedy category* is a small category $\mathbb{R}$ together with subcategories (on the same objects) $\mathbb{R}^{+}$ and $\mathbb{R}^{-}$ and a degree function $d\colon \mathrm{Ob}(\mathbb{R}) \longrightarrow \mathbb{N}$ such that:

(i) Every non-identity morphism in $\mathbb{R}^{+}$ raises degree and every non-identity morphism in $\mathbb{R}^{-}$ lowers degree.

(ii) Every morphism $r \xrightarrow{f} s$ in $\mathbb{R}$ factors uniquely as $r \xrightarrow{h} t \xrightarrow{g} s$, where $h \in \mathbb{R}^{-}$ and $g \in \mathbb{R}^{+}$.

Any strict Reedy category is dualizable, i.e., if $\mathbb{R}$ is a strict Reedy category, so is its opposite $\mathbb{R}^{\mathrm{op}}$, by switching the roles of $\mathbb{R}^{+}$ and $\mathbb{R}^{-}$. However, the notion

I. Moerdijk and B. Toën, *Simplicial Methods for Operads and Algebraic Geometry*, Advanced Courses in Mathematics - CRM Barcelona, DOI 10.1007/978-3-0348-0052-5_5, © Springer Basel AG 2010

of a strict Reedy category is not invariant under equivalence of categories. In fact, in a strict Reedy category every automorphism is an identity.

*Example* 5.1.2. The following are examples of strict Reedy categories.

(i) The simplicial category $\Delta$. The degree function is defined by $d([n]) = n$; the subcategory $\Delta^+$ consists of the monomorphisms (compositions of faces), and the subcategory $\Delta^-$ consists of the epimorphisms (compositions of degeneracies).

(ii) The category $\Delta^{\mathrm{op}}$.

(iii) The category of natural numbers $(\mathbb{N}, \leq)$.

(iv) The categories $\cdot \longleftarrow \cdot \longrightarrow \cdot$ and $\cdot \rightrightarrows \cdot$.

(v) The category of planar trees $\Omega_p$. The degree function is given by the number of vertices in the tree. The morphisms in $\Omega_p^+$ are the ones inducing an injection between the sets of edges, and the morphisms in $\Omega_p^-$ are the ones inducing a surjection between the sets of edges.

We recall the definition of latching and matching objects in the functor category $\mathcal{E}^{\mathbb{R}}$. Let $X \colon \mathbb{R} \longrightarrow \mathcal{E}$ be a functor and $r$ an object of $\mathbb{R}$. The *r-th latching object* $L_r(X)$ is defined to be

$$L_r(X) = \varinjlim_{s \to r} X_s$$

where the colimit is taken over the full subcategory of $\mathbb{R}^+/r$ excluding the identity morphism. Dually, the *r-th matching object* $M_r(X)$ is defined to be

$$M_r(X) = \varprojlim_{r \to s} X_s$$

where the limit is taken over the full subcategory $r/\mathbb{R}^-$ excluding the identity morphism. We will assume that the category $\mathcal{E}$ is sufficiently complete and cocomplete for these limits and colimits to exist.

For every $X$ in $\mathcal{E}^{\mathbb{R}}$ and every $r$ in $\mathbb{R}$ there are natural maps

$$L_r(X) \longrightarrow X_r \longrightarrow M_r(X)$$

relating the latching and matching objects.

## 5.2 Model structures for strict Reedy categories

Let $\mathcal{E}$ be a cofibrantly generated model category and $\mathbb{R}$ a strict Reedy category. The model structure on the functor category $\mathcal{E}^{\mathbb{R}}$ is defined in terms of the latching and matching objects.

**Definition 5.2.1.** Let $f\colon X \longrightarrow Y$ be a morphism in $\mathcal{E}^{\mathbb{R}}$.

(i) $f$ is a *Reedy weak equivalence* if, for each $r$ in $\mathbb{R}$, the induced map

$$f_r\colon X_r \longrightarrow Y_r$$

is a weak equivalence in $\mathcal{E}$.

(ii) $f$ is a *Reedy fibration* if, for each $r$ in $\mathbb{R}$, the natural map

$$X_r \longrightarrow M_r(X) \times_{M_r(Y)} Y_r$$

is a fibration in $\mathcal{E}$.

(iii) $f$ is a *Reedy cofibration* if, for each $r$ in $\mathbb{R}$, the natural map

$$X_r \cup_{L_r(X)} L_r(Y) \longrightarrow Y_r$$

is a cofibration in $\mathcal{E}$.

The following classical result provides the functor category $\mathcal{E}^{\mathbb{R}}$ with a model structure. In the case of the category of cosimplicial spaces, this fact was proved by Bousfield and Kan [BK72]. A model structure for simplicial objects in arbitrary model categories was proved by Reedy [Ree74]. An exposition of these results with more details can be found in [Hir03, Ch. 15] or [Hov99, §5.2].

**Theorem 5.2.2.** *Let $\mathcal{E}$ be a cofibrantly generated model category and let $\mathbb{R}$ be a strict Reedy category. Then the functor category $\mathcal{E}^{\mathbb{R}}$ has a (so-called Reedy) model structure in which the weak equivalences, fibrations and cofibrations are the Reedy weak equivalences, Reedy fibrations and Reedy cofibrations respectively, described above.* $\square$

## 5.3   Generalized Reedy categories

An important fact about strict Reedy categories is that every automorphism is an identity. In this section we introduce the notion of a generalized Reedy category extending that of strict Reedy categories, by allowing non-trivial automorphisms.

**Definition 5.3.1.** A *generalized Reedy category* is a small category $\mathbb{R}$ together with subcategories (on the same objects) $\mathbb{R}^+$ and $\mathbb{R}^-$ and a degree function $d\colon \mathrm{Ob}(\mathbb{R}) \longrightarrow \mathbb{N}$ such that:

(i) Every non-isomorphism in $\mathbb{R}^+$ raises degree; every non-isomorphism in $\mathbb{R}^-$ lowers degree; every isomorphism in $\mathbb{R}$ preserves the degree.

(ii) $\mathbb{R}^+ \cap \mathbb{R}^- = \mathrm{Iso}(\mathbb{R})$, where $\mathrm{Iso}(\mathbb{R})$ denotes the set of isomorphisms in $\mathbb{R}$.

(iii) Every morphism $f$ in $\mathbb{R}$ factors as $f = g \circ h$ with $g \in \mathbb{R}^+$ and $h \in \mathbb{R}^-$, and this factorization is unique up to isomorphism.

(iv) If $\theta \circ f = f$ and $\theta$ is an isomorphism in $\mathbb{R}$ and $f \in \mathbb{R}^-$, then $\theta$ is an identity.

We say that $\mathbb{R}$ is *dualizable* if, moreover, the following condition holds:

(iv)' If $f \circ \theta = f$ and $\theta$ is an isomorphism in $\mathbb{R}$ and $f \in \mathbb{R}^+$, then $\theta$ is an identity.

*Remark* 5.3.2. The notion of generalized Reedy category is invariant under equivalence of categories, i.e., if $\mathbb{R} \longrightarrow \mathbb{R}'$ is an equivalence of categories and $\mathbb{R}$ is a generalized Reedy category, so is $\mathbb{R}'$. Note that a generalized Reedy category $\mathbb{R}$ is dualizable if and only if the opposite category $\mathbb{R}^{\mathrm{op}}$ is a generalized Reedy category.

*Example* 5.3.3. The following are examples of dualizable generalized Reedy categories:

(i) Any strict Reedy category is a generalized Reedy category. A generalized Reedy category is equivalent to a strict one if and only if it has no non-trivial automorphisms, and is itself strict if and only if it is moreover skeletal.

(ii) The cyclic category $\Lambda$.

(iii) Segal's category $\Gamma$ is a generalized Reedy category. In fact, $\Gamma^{\mathrm{op}}$ is equivalent to the category $Fin_*$ of finite pointed sets, and one can take $Fin_*^+$ to consist of monomorphisms and $Fin_*^-$ of epimorphisms, while the degree function is given by cardinality.

(iv) The category of finite sets $Fin$.

(v) The category of trees $\Omega$. The degree function is given by the number of vertices in the tree. The morphisms in $\Omega^+$ are the ones inducing an injection between the sets of edges, and the morphisms in $\Omega^-$ are the ones inducing a surjection between the sets of edges.

For a generalized Reedy category, the definition of latching and matching objects still makes sense. These notions generalize the notions of latching and matching objects in the case of strict Reedy categories.

Let $\mathbb{R}$ be a generalized Reedy category and $r$ any object of $\mathbb{R}$. We denote by $\mathbb{R}^+(r)$ the full subcategory of $\mathbb{R}^+/r$ excluding the invertible morphisms of $\mathbb{R}^+$ with codomain $r$. Similarly, we denote by $\mathbb{R}^-(r)$ the full subcategory of $r/\mathbb{R}^-$ excluding the invertible morphisms of $\mathbb{R}^-$ with domain $r$.

**Definition 5.3.4.** Let $\mathbb{R}$ be a generalized Reedy category. Let $r$ be an object of $\mathbb{R}$ and $X$ and object of $\mathcal{E}^{\mathbb{R}}$.

(i) The *$r$-th latching object* $L_r(X)$ is defined to be

$$L_r(X) = \varinjlim_{s \to r} X_s$$

where the colimit is taken over the category $\mathbb{R}^+(r)$.

(ii) The *r-th matching object* $M_r(X)$ is defined to be

$$M_r(X) = \varprojlim_{r \to s} X_s$$

where the limit is taken over the category $\mathbb{R}^-(r)$.

We will assume that the category $\mathcal{E}$ is sufficiently complete and cocomplete for these limits and colimits to exist.

Note that, for every $r \in \mathbb{R}$, we have that $\mathrm{Aut}(r) = \mathrm{Iso}_{\mathbb{R}}(r, r)$ acts both on $L_r(X)$ and $M_r(X)$ and that there are natural $\mathrm{Aut}(r)$-equivariant morphisms

$$L_r(X) \longrightarrow X_r \longrightarrow M_r(X).$$

It is possible to give a more global definition of latching and matching objects by using groupoids of objects of fixed degree. For each natural number $n$, the full subgroupoid of $\mathrm{Iso}(\mathbb{R})$ spanned by the objects of degree $n$ will be denoted by $\mathbb{G}_n(\mathbb{R})$. We denote by $\mathbb{R}^+((n))$ the category with objects the non-invertible morphisms $u \colon s \longrightarrow r$ in $\mathbb{R}^+$ such that $d(r) = n$, and with morphisms from $u$ to $u'$ the commutative squares

$$
\begin{array}{ccc}
s & \xrightarrow{\ f\ } & s' \\
{\scriptstyle u}\big\downarrow & & \big\downarrow{\scriptstyle u'} \\
r & \xrightarrow[\ g\ ]{\sim} & r'
\end{array}
$$

such that $f \in \mathbb{R}^+$ and $g \in \mathbb{G}_n(\mathbb{R})$. The subcategory $\mathbb{R}^+(n)$ of $\mathbb{R}^+((n))$ contains all the morphisms for which $g$ is an identity. Observe that

$$\mathbb{R}^+(n) = \coprod_{d(r)=n} \mathbb{R}^+(r),$$

where $\mathbb{R}^+(r)$ is the category used in Definition 5.3.4. There is a diagram of categories

$$\mathbb{R} \xleftarrow{\ d_n\ } \mathbb{R}^+((n)) \xrightarrow{\ c_n\ } \mathbb{G}_n(\mathbb{R}) \xrightarrow{\ j_n\ } \mathbb{R}$$

where $d_n$ is the domain functor, $c_n$ is the codomain functor and $j_n$ is the inclusion.

Dually, we denote by $\mathbb{R}^-((n))$ the category with objects the non-invertible morphisms $u \colon r \longrightarrow s$ in $\mathbb{R}^-$ such that $d(r) = n$, and with morphisms from $u$ to $u'$ the commutative squares

$$
\begin{array}{ccc}
r & \xrightarrow[\sim]{\ g\ } & r' \\
{\scriptstyle u}\big\downarrow & & \big\downarrow{\scriptstyle u'} \\
s & \xrightarrow[\ f\ ]{} & s'
\end{array}
$$

such that $f \in \mathbb{R}^-$ and $g \in \mathbb{G}_n(\mathbb{R})$. The subcategory $\mathbb{R}^-(n)$ of $\mathbb{R}^-((n))$ contains all the morphisms for which $g$ is an identity. Observe that

$$\mathbb{R}^-(n) = \coprod_{d(r)=n} \mathbb{R}^-(r).$$

There is a diagram of categories

$$\mathbb{R} \xleftarrow{\gamma_n} \mathbb{R}^-((n)) \xrightarrow{\delta_n} \mathbb{G}_n(\mathbb{R}) \xrightarrow{j_n} \mathbb{R}$$

where $\gamma_n$ is the codomain functor, $\delta_n$ is the domain functor and $j_n$ is the inclusion.

The definition of a *latching object* $L_n(X)$ and a *matching object* $M_n(X)$ in $\mathcal{E}^{\mathbb{G}_n(\mathbb{R})}$ for an object $X$ of $\mathcal{E}^{\mathbb{R}}$ is now the following:

$$L_n(X) = (c_n)_! d_n^*(X) \quad \text{and} \quad M_n(X) = (\delta_n)_* \gamma_n^*(X).$$

Here we have used the following notation: for a functor $f \colon \mathbb{A} \longrightarrow \mathbb{B}$ between small categories, we denote by $f^* \colon \mathcal{E}^{\mathbb{B}} \longrightarrow \mathcal{E}^{\mathbb{A}}$ the evident functor given by reindexing the diagrams, while $f_!$, $f_* \colon \mathcal{E}^{\mathbb{A}} \longrightarrow \mathcal{E}^{\mathbb{B}}$ are its left and right adjoints.

We write $X_n = j_n^*(X)$, so that in each degree $n$ we have a *latching map* and a *matching map*

$$L_n(X) \longrightarrow X_n \longrightarrow M_n(X).$$

Observe that

$$L_n(X)_r = \varinjlim_{s \to r} X_s \quad \text{and} \quad M_n(X)_r = \varprojlim_{r \to s} X_s$$

where the colimit is taken over the category $\mathbb{R}^+(r)$ and the limit is taken over the category $\mathbb{R}^-(r)$. Thus, we will simplify the notation and write $L_r(X)$ for $L_n(X)_r$ and $M_r(X)$ for $M_n(X)_r$.

## 5.4 Model structures for generalized Reedy categories

Recall that if $\mathcal{E}$ is a cofibrantly generated model category and $G$ is a finite group, then there is a model structure in the category $\mathcal{E}^G$ of objects of $\mathcal{E}$ with a right $G$-action, in which weak equivalences and fibrations are defined pointwise, i.e., by forgetting the $G$-action.

Let $\mathcal{E}$ be a cofibrantly generated model category and $\mathbb{R}$ a generalized Reedy category. The model structure on the functor category $\mathcal{E}^{\mathbb{R}}$ is defined in terms of latching and matching objects and the model structure of the category $\mathcal{E}^{\mathrm{Aut}(r)}$ for every $r$ in $\mathbb{R}$.

**Definition 5.4.1.** Let $f \colon X \longrightarrow Y$ be a morphism in $\mathcal{E}^{\mathbb{R}}$.

(i) $f$ is a *Reedy weak equivalence* if, for each $r$ in $\mathbb{R}$, the induced map

$$f_r \colon X_r \longrightarrow Y_r$$

is a weak equivalence in $\mathcal{E}^{\mathrm{Aut}(r)}$ (i.e., a weak equivalence in $\mathcal{E}$).

(ii) $f$ is a *Reedy fibration* if, for each $r$ in $\mathbb{R}$, the natural map

$$X_r \longrightarrow M_r(X) \times_{M_r(Y)} Y_r$$

is a fibration in $\mathcal{E}^{\mathrm{Aut}(r)}$ (i.e., a fibration in $\mathcal{E}$).

(iii) $f$ is a *Reedy cofibration* if, for each $r$ in $\mathbb{R}$, the natural map

$$X_r \cup_{L_r(X)} L_r(Y) \longrightarrow Y_r$$

is a cofibration in $\mathcal{E}^{\mathrm{Aut}(r)}$.

A Reedy fibration that is also a Reedy weak equivalence is called a *Reedy trivial fibration*. A Reedy cofibration that is also a Reedy weak equivalence is called a *Reedy trivial cofibration*.

We can reformulate the definition of the classes of maps in Definition 5.4.1 by using the latching and matching objects defined previously.

**Lemma 5.4.2.** *Let $\mathcal{E}$ be a cofibrantly generated model category, $\mathbb{R}$ a generalized Reedy category, and $f: X \longrightarrow Y$ a morphism in $\mathcal{E}^{\mathbb{R}}$.*

(i) *$f$ is a Reedy weak equivalence if and only if, for each natural number $n$, the morphism $X_n \longrightarrow Y_n$ is a weak equivalence in $\mathcal{E}^{\mathbb{G}_n(\mathbb{R})}$.*

(ii) *$f$ is a Reedy fibration if and only if, for each natural number $n$, the morphism $X_n \longrightarrow M_n(X) \times_{M_n(Y)} Y_n$ is a fibration in $\mathcal{E}^{\mathbb{G}_n(\mathbb{R})}$.*

(iii) *$f$ is a Reedy cofibration if and only if, for each natural number $n$, the morphism $X_n \cup_{L_n(X)} L_n(Y) \longrightarrow Y_n$ is a cofibration in $\mathcal{E}^{\mathbb{G}_n(\mathbb{R})}$.*

*Proof.* There is an equivalence of categories $\mathcal{E}^{\mathbb{G}_n(\mathbb{R})} \xrightarrow{\sim} \prod_r \mathcal{E}^{\mathrm{Aut}(r)}$ where $r$ runs through a set of representatives for the connected components of the groupoid $\mathbb{G}_n(\mathbb{R})$. $\qquad\square$

To prove the main theorem of this section we will need some technical lemmas on Reedy fibrations and cofibrations. We omit the proofs, that can be found in [BM08, §5].

**Lemma 5.4.3.** *Let $f: A \longrightarrow B$ be a Reedy cofibration such that $f_r: A_r \longrightarrow B_r$ is a weak equivalence for all objects $r$ of $\mathbb{R}$ of degree $< n$. Then the induced map $L_n(f): L_n(A) \longrightarrow L_n(B)$ is a pointwise trivial cofibration (i.e., $L_n(f)_r$ is a trivial cofibration in $\mathcal{E}$ for each object $r$ of $\mathbb{R}$). If $f$ is a Reedy trivial cofibration, then it has the left lifting property with respect to Reedy fibrations.* $\qquad\square$

**Lemma 5.4.4.** *Let $f: A \longrightarrow B$ be a Reedy fibration such that $f_r: A_r \longrightarrow B_r$ is a weak equivalence for all objects $r$ of $\mathbb{R}$ of degree $< n$. Then the induced map $M_n(f): M_n(A) \longrightarrow M_n(B)$ is a pointwise trivial fibration (i.e., $M_n(f)_r$ is a trivial fibration in $\mathcal{E}$ for each object $r$ of $\mathbb{R}$). If $f$ is a Reedy trivial fibration, then it has the right lifting property with respect to Reedy cofibrations.* $\qquad\square$

The following theorem provides a model structure for the functor category $\mathcal{E}^{\mathbb{R}}$ when $\mathbb{R}$ is a generalized Reedy category.

**Theorem 5.4.5.** *Let $\mathbb{R}$ be a generalized Reedy category and $\mathcal{E}$ a cofibrantly generated model category. Then the functor category $\mathcal{E}^{\mathbb{R}}$ with the Reedy weak equivalences, Reedy fibrations and Reedy cofibrations is a model category.*

*Proof.* The category $\mathcal{E}^{\mathbb{R}}$ is complete and cocomplete. Limits and colimits in $\mathcal{E}^{\mathbb{R}}$ are constructed pointwise. The class of Reedy weak equivalences has the 'two out of three' property. Moreover, all three classes are closed under retracts. The lifting axiom is proved in Lemma 5.4.3 and Lemma 5.4.4.

It only remains to prove the factorization axiom. Given a map $f \colon X \longrightarrow Y$ in $\mathcal{E}^{\mathbb{R}}$, we are going to construct inductively a factorization $X \longrightarrow A \longrightarrow Y$ of $f$ into a trivial Reedy cofibration followed by a Reedy fibration.

For $n = 0$, factor $f_0$ in $\mathcal{E}^{\mathbb{G}_0(\mathbb{R})}$ as $X_0 \longrightarrow A_0 \longrightarrow Y_0$ into a trivial cofibration followed by a fibration. Next, if $X_{\leq n-1} \longrightarrow A_{\leq n-1} \longrightarrow Y_{\leq n-1}$ is a factorization of $f_{\leq n-1}$ into a trivial Reedy cofibration followed by a Reedy fibration in $\mathcal{E}^{\mathbb{R}_{\leq n-1}}$, where $\mathbb{R}_{\leq n-1}$ is the full subcategory of $\mathbb{R}$ of objects of degree $\leq n-1$, we obtain the following commutative diagram in $\mathcal{E}^{\mathbb{G}_n(\mathbb{R})}$:

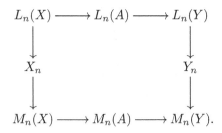

In this diagram, $L_n(A)$ denotes the object defined by the usual formula

$$L_n(A) = \varinjlim_{s \to r} A_s,$$

which makes sense although, up to now, $A$ has only been defined on $\mathbb{R}_{\leq n-1}$. Similarly for $M_n(A)$. There is a again a canonical map $L_n(A) \longrightarrow M_n(A)$ fitting in the diagram above. Together with the given map $X_n \longrightarrow Y_n$, this induces a map $X_n \cup_{L_n(X)} L_n(A) \longrightarrow M_n(A) \times_{M_n(Y)} Y_n$ which we factor as a trivial cofibration followed by a fibration in $\mathcal{E}^{\mathbb{G}_n(\mathbb{R})}$:

$$X_n \cup_{L_n(X)} L_n(A) \overset{\sim}{\rightarrowtail} A_n \twoheadrightarrow M_n(A) \times_{M_n(Y)} Y_n.$$

The object $A_n$ of $\mathcal{E}^{\mathbb{G}_n(\mathbb{R})}$ together with the maps $L_n(A) \longrightarrow A_n \longrightarrow M_n(A)$ define an extension of $A_{\leq n-1}$ to an object $A_{\leq n}$ in $\mathcal{E}^{\mathbb{G}_n(\mathbb{R})}$ together with a factorization of $f_{\leq n} \colon X_{\leq n} \longrightarrow Y_{\leq n}$ into a Reedy cofibration $X_{\leq n} \longrightarrow A_{\leq n}$ followed by a Reedy fibration $A_{\leq n} \longrightarrow Y_{\leq n}$. The former map is a trivial Reedy cofibration, because the map $X_n \longrightarrow A_n$ decomposes into two maps $X_n \longrightarrow X_n \cup_{L_n(X)} L_n(A) \longrightarrow A_n$, the first one of which is a weak equivalence by Lemma 5.4.3, and the second one by construction. This defines the required factorization of $f_{\leq n}$ in $\mathcal{E}^{\mathbb{R}_{\leq n}}$.

The factorization of $f$ into a Reedy cofibration followed by a trivial Reedy fibration is constructed dually, using Lemma 5.4.4 instead of Lemma 5.4.3.  □

## 5.5  Dendroidal objects and simplicial objects

The categories $\Delta^{\mathrm{op}}$, $\Omega_p^{\mathrm{op}}$ and $\Omega^{\mathrm{op}}$ are generalized Reedy categories, as we have seen in Example 5.3.3. We will write $s\mathcal{E} = \mathcal{E}^{\Delta^{\mathrm{op}}}$ for the category of simplicial objects in $\mathcal{E}$, $pd\mathcal{E} = \mathcal{E}^{\Omega_p^{\mathrm{op}}}$ for the category of planar dendroidal objects in $\mathcal{E}$, and $d\mathcal{E} = \mathcal{E}^{\Omega^{\mathrm{op}}}$ for the category of dendroidal objects in $\mathcal{E}$. If $\mathcal{E}$ is a cofibrantly generated model category, then there is a model structure in $s\mathcal{E}$, $pd\mathcal{E}$ and $d\mathcal{E}$ described in Theorem 5.4.5. The inclusion functors

induce the corresponding diagrams of adjoint functors

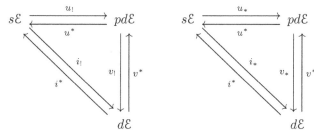

where $u_!$, $v_!$ and $i_!$ are left adjoints to $u^*$, $v^*$ and $i^*$, and $u_*$, $v_*$ and $i_*$ are right adjoints to $u^*$, $v^*$ and $i^*$ respectively.

We will show that these functors are Quillen pairs between the Reedy model structures of simplicial objects in $\mathcal{E}$ and dendroidal objects in $\mathcal{E}$.

**Lemma 5.5.1.** *The functor $i^*\colon d\mathcal{E} \longrightarrow s\mathcal{E}$ preserves Reedy weak equivalences.*

*Proof.* Let $f\colon X \longrightarrow Y$ be a Reedy weak equivalence in $d\mathcal{E}$. This means that $f_T\colon X_T \longrightarrow Y_T$ is a weak equivalence in $\mathcal{E}^{\mathrm{Aut}(T)}$ for every tree $T$ in $\Omega$. In particular, it is a weak equivalence when $T = [n]$, the linear tree with $n$ vertices for every $n$. Therefore $i^*(f)_{[n]}\colon X_{[n]} \longrightarrow Y_{[n]}$ is a weak equivalence for every $n$ and hence $i^*(f)$ is a weak equivalence in $s\mathcal{E}$.  □

In order to prove that $i^*$ preserves fibrations, we need to establish the relation between matching objects in $d\mathcal{E}$ and matching objects in $s\mathcal{E}$.

**Lemma 5.5.2.** $M_{[n]}(i^*(X)) = M_{i([n])}(X)$ *for every $X$ in $d\mathcal{E}$ and every $n$.*

*Proof.* We know that $M_{i([n])}X = \varprojlim_{T \to i([n])} X_T$. Since the maps $T \longrightarrow i([n])$ used to define the limit are in $\Omega^+$ (i.e., are compositions of face maps), this implies that $T = i([m])$ for some $m < n$. It follows that

$$M_{i([n])}X = \varprojlim_{i([m]) \to i([n])} X_{i([m])} = \varprojlim_{[m] \to [n]} X(i([m])) = M_{[n]}(i^*(X))$$

for every $n$. □

**Proposition 5.5.3.** *The pair of adjoint functors $(i_!, i^*)$ is a Quillen pair.*

*Proof.* The functor $i^*$ preserves Reedy fibrations by Lemma 5.5.2 and trivial Reedy fibrations. Hence, its left adjoint $i_!$ preserves Reedy cofibrations and trivial Reedy cofibrations. □

The functor $i^*$ preserves not only fibrations but also cofibrations:

**Lemma 5.5.4.** $L_{[n]}(i^*X) = L_{i([n])}X$ *for every $X$ in $d\mathcal{E}$ and every $n$.*

*Proof.* We know that $L_{i([n])}X = \varinjlim_{i([n]) \to T} X_T$. Since $i([n])$ is a linear tree and the maps $i([n]) \longrightarrow T$ are compositions of degeneracies, it implies that $T$ has to be a linear tree $T = i([m])$ for some $m < n$. It follows that

$$L_{i([n])}X = \varinjlim_{[n] \to [m]} i^*X = L_{[n]}i^*X$$

for every $n$. □

**Proposition 5.5.5.** *The pair of adjoint functors $(i^*, i_*)$ is a Quillen pair.*

*Proof.* The functor $i^*$ preserves Reedy cofibrations by Lemma 5.5.4 and trivial Reedy cofibrations. Hence, its right adjoint $i_*$ preserves Reedy fibrations and trivial Reedy fibrations. □

A similar argument for the inclusion $u\colon \Delta \longrightarrow \Omega_p$ gives the following proposition.

**Proposition 5.5.6.** *The pairs of adjoint functors $(u_!, u^*)$ and $(u^*, u_*)$ are Quillen pairs.* □

If we consider the inclusion $v\colon \Omega_p \longrightarrow \Omega$, one can use the fact that any map in $\Omega^+$ and $\Omega^-$ factors as an isomorphism followed by a planar map to prove analogues to Lemma 5.5.2 and Lemma 5.5.4:

**Lemma 5.5.7.** $M_T(v_*X) = M_{v(T)}X$ *and* $L_T(v^*X) = L_{v(T)}X$ *for every $X$ in $d\mathcal{E}$ and every $T$ in $\Omega$.* □

**Corollary 5.5.8.** *The pairs of adjoint functors* $(v_!, v^*)$ *and* $(v^*, v_*)$ *are Quillen pairs.*

*Proof.* The functor $v^*$ preserves Reedy weak equivalences, since they are defined levelwise. By Lemma 5.5.7, it preserves Reedy fibrations, trivial Reedy fibrations, Reedy cofibrations and trivial Reedy cofibrations. Hence, its left adjoint $v_!$ preserves Reedy cofibrations and trivial Reedy cofibrations, and its right adjoint $i_*$ preserves Reedy fibrations and trivial Reedy fibrations. □

## 5.6  Dendroidal Segal objects

In this section we introduce the notion of a (complete) dendroidal Segal object and prove that there is a model structure on the category of dendroidal spaces whose fibrant objects are the (complete) dendroidal Segal spaces. We will see later that the model structure of complete dendroidal Segal spaces is Quillen equivalent to the model structure for dendroidal sets given in Lecture 8. A more detailed discussion of the subject of this section can be found in [CM09b].

Let $T$ be any object of $\Omega$ with at least one vertex. The *Segal core* of $\Omega[T]$ is the subobject

$$\mathrm{Sc}[T] \rightarrowtail \Omega[T]$$

given by all the corollas in $T$; more precisely, $\mathrm{Sc}[T]$ is the union of all the images of morphisms $\Omega[R] \rightarrowtail \Omega[T]$ where $R$ is a corolla and $R \rightarrowtail T$ is completely determined by the vertex of $T$ in its image. If we denote by $C_n$ the $n$-th corolla

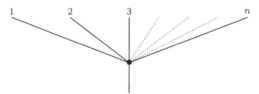

then we can write

$$\mathrm{Sc}[T] = \bigcup_v \Omega[C_{n(v)}],$$

where the union is over all the vertices $v$ in $T$, and $n(v)$ is the number of input edges at $v$. It will be convenient to define $\mathrm{Sc}[T] = \Omega[T]$ if $T$ is the tree with no vertices; i.e., $\Omega[T] = U$.

In the following definition, $J$ denotes the dendroidal object obtained from the category $J = (0 \xleftrightarrow{\sim} 1)$ by including it into dendroidal sets and then applying the functor $dSets \longrightarrow d\mathcal{E}$ induced by the strong monoidal functor $Sets \longrightarrow \mathcal{E}$, as in Example 1.3.6. Similarly, $\Omega[T]$ and $\mathrm{Sc}[T]$ are viewed as objects of $\mathcal{E}^{\Omega^{\mathrm{op}}}$ in (5.1). The category $\mathcal{E}^{\Omega^{\mathrm{op}}}$ is closed monoidal (see Remark 4.2.6) with internal hom denoted by $\mathrm{Hom}(-, -)$, while $\mathrm{hom}(-, -)$ is the $\mathcal{E}$-valued hom defined by $\mathrm{hom}(X, Y) = \mathrm{Hom}(X, Y)_T$ for $T = \eta$.

**Definition 5.6.1.** Let $\mathcal{E}$ be a cofibrantly generated monoidal model category and $X$ a dendroidal object in $\mathcal{E}$. Then $X$ is said to satisfy the *Segal condition* if, for any tree $T$, the map

$$\hom(\Omega[T], X) \longrightarrow \hom(\mathrm{Sc}[T], X) \tag{5.1}$$

is a weak equivalence in $\mathcal{E}$. If moreover the map

$$\mathrm{Hom}(J, X) \longrightarrow \mathrm{Hom}(\{0\}, X) = X \tag{5.2}$$

is a weak equivalence in $\mathcal{E}^{\Omega^{\mathrm{op}}}$, then $X$ is said to satisfy the *complete Segal condition*.

A Reedy fibrant dendroidal object satisfying the Segal condition will be called a *dendroidal Segal object*. A Reedy fibrant dendroidal object satisfying the complete Segal condition will be called a *complete dendroidal Segal object*.

*Remark 5.6.2.* If $\mathcal{E}$ is the category of simplicial sets and $X$ is a Reedy fibrant dendroidal space, then the map (5.1) is a Kan fibration of simplicial sets. So a Reedy fibrant dendroidal set is a dendroidal Segal space if and only if the map (5.1) is a trivial fibration of simplicial sets, for any tree $T$. Thus, the following are equivalent for a Reedy fibrant $X$ (recall that an object $X$ is said to have the right lifting property with respect to a map $A \longrightarrow B$ if the map $X \longrightarrow 1$ does):

(i) $X$ is a dendroidal Segal space.

(ii) $X$ has the right lifting property with respect to

$$\partial\Delta[n] \otimes \Omega[T] \cup \Delta[n] \otimes \mathrm{Sc}[T] \rightarrowtail \Delta[n] \otimes \Omega[T]$$

for any $n \geq 0$ and any tree $T$ with at least one vertex.

(iii) $X$ has the right lifting property with respect to

$$A \otimes \Omega[T] \cup B \otimes \mathrm{Sc}[T] \rightarrowtail B \otimes \Omega[T]$$

for any monomorphism $A \rightarrowtail B$ of simplicial sets and any tree $T$ with at least one vertex.

One can obtain a similar characterization for complete dendroidal Segal spaces by adding the map (5.2) to the maps $\mathrm{Sc}[T] \rightarrowtail \Omega[T]$ in the right lifting property conditions.

**Theorem 5.6.3.** *There is a model structure on the category of dendroidal spaces, called the (complete) Segal model structure, in which the fibrant objects are precisely the (complete) Segal dendroidal spaces.*

*Proof.* The Segal model structure is obtained by a left Bousfield localization of the Reedy model structure on $sSets^{\Omega^{\mathrm{op}}}$ with respect to the map

$$\coprod_T (\mathrm{Sc}[T] \rightarrowtail \Omega[T]),$$

where $T$ runs over all the trees with at least one vertex. The complete Segal model structure is obtained from the previous one by localizing further with respect to the map $\{0\} \longrightarrow J$. □

The 'complete' model structure of Theorem 5.6.3 is closely related to the model structure on dendroidal sets to be discussed in Lecture 8. In fact, one has the following theorem, for the proof of which we refer to [CM09b]:

**Theorem 5.6.4.** *The category of dendroidal spaces with the complete Segal model structure is Quillen equivalent to the category of dendroidal sets.* □

# Lecture 6

# Boardman–Vogt resolution and homotopy coherent nerve

In this lecture, we describe a generalization of the $W$-construction of Boardman and Vogt for coloured operads in any monoidal category with a suitable notion of interval. This generalized $W$-construction is then used to define the homotopy coherent dendroidal nerve of a given coloured operad $P$. The homotopy coherent nerve will play a fundamental role in the definition of homotopy $P$-algebras and weak higher categories.

## 6.1 The classical $W$-construction

In this section we recall the Boardman–Vogt resolution of operads in the category of topological spaces (see [BV73, Ch. III]).

Let $Top$ denote the category of compactly generated topological spaces. The $W$-construction is a functor

$$W\colon Oper_C(Top) \longrightarrow Oper_C(Top),$$

where $Oper_C(Top)$ denotes the category of $C$-coloured operads in $Top$, together with a natural transformation $\gamma\colon W \longrightarrow \mathrm{id}$ (i.e., $W$ is an augmented endofunctor in $Oper_C(Top)$). To each topological coloured operad $P$ there is an associated topological coloured operad $W(P)$ and a natural map $\gamma_P\colon W(P) \longrightarrow P$.

If we think of the maps of coloured operads $P \longrightarrow Top$ (considering $Top$ as a coloured operad with the cartesian product) as describing $P$-algebras in $Top$, then the maps $W(P) \longrightarrow Top$ will describe *homotopy $P$-algebras* in $Top$. The augmentation induces a map

$$Oper_C(Top)(P, Top) \xrightarrow{\gamma_P^*} Oper_C(Top)(W(P), Top)$$

I. Moerdijk and B. Toën, *Simplicial Methods for Operads and Algebraic Geometry*, Advanced Courses in Mathematics - CRM Barcelona, DOI 10.1007/978-3-0348-0052-5_6, © Springer Basel AG 2010

which views any $P$-algebra as a homotopy $P$-algebra.

In the case of non-symmetric operads, the functor $W$ can be explicitly described as follows. Let $P$ be a non-symmetric $C$-coloured operad in *Top* and let $H = [0, 1]$ be the unit interval. The colours of $W(P)$ are the same as those of $P$ and the space of operations $W(P)(c_1, \ldots, c_n; c)$ is a quotient of a space of labelled planar trees. We consider, for $c_1, \ldots, c_n, c$ in $C$, the topological space $A(c_1, \ldots, c_n; c)$ of planar trees whose edges are labelled by elements of $C$ and, in particular, for every such tree the input edges are labelled by the given $c_1, \ldots, c_n$ and the output edge is labelled by $c$. We assign to each of the inner edges of these trees a length $t \in H$. Each vertex with input edges labelled by $b_1, \ldots, b_m \in C$ (in the planar order) and output edge labelled by $b \in C$ is labelled by an element of $P(b_1, \ldots, b_m; b)$.

*Example* 6.1.1. The following tree is an element of $A(c, c, d; e)$:

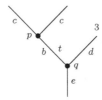

where $p \in P(c, c; b)$, $q \in P(b, d; e)$, and $t \in [0, 1]$.

There is a canonical topology in $A(c_1, \ldots, c_n; c)$ induced by the topology of $P$ and the standard topology of the unit interval. The space $W(P)(c_1, \ldots, c_n; c)$ is the quotient space of $A(c_1, \ldots, c_n; c)$ obtained by the following two relations:

(i) If a tree has a unary vertex $v$ labelled by an identity, then we identify such tree with the tree obtained by removing this vertex and identifying the input edge $x$ of $v$ with its output edge $y$. The length assigned to the new edge is the maximum of the lengths of the edges $x$ and $y$, or it has no length if the new edge is outer.

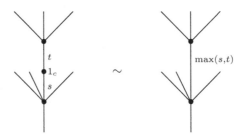

(ii) If there is a tree with an internal edge $e$ with zero length, then we identify it with the tree obtained by contracting the edge $e$, using the corresponding $\circ_i$

operation of the coloured operad $P$.

The collection $W(P)(c_1, \ldots, c_n; c)$ for $c_1, \ldots, c_n, c \in C$ forms a $C$-coloured operad. The unit for each colour $c$ is the tree | coloured by $c$. Composition is given by grafting, assigning length 1 to the newly arisen internal edges.

*Remark* 6.1.2. There is a $W$-construction for symmetric operads defined in a similar way (see [BM07, §3]). The forgetful functor from symmetric operads to non-symmetric operads has a left adjoint which identifies the category of non-symmetric operads with a full coreflective subcategory of the category of symmetric operads. If

$$\Sigma : Oper_\Sigma(Top) \rightleftarrows Oper(Top) : U$$

is the free-forgetful adjunction relating non-symmetric operads and symmetric operads, then $W(\Sigma P) = \Sigma(WP)$.

*Example* 6.1.3. Let $P$ be the non-symmetric coloured operad with only one colour and $P(c, \overset{(n)}{\ldots}, c; c) = P(n)$ consisting of a single $n$-ary operation for $n \geq 1$ and $P(\ ; c) = P(0) = \emptyset$.

The operad $W(P)$ will again have only one colour. Since every unary vertex in a labelled tree in $W(P)(n)$ can only be labelled by the identity, it is enough (by relation (i) in the definition of $W(P)$) to consider only those trees in $W(P)(n)$ without unary vertices. We call these trees *regular trees*. Thus, if $n = 1, 2$, then $W(P)(n)$ is a one-point space. For the case $n = 3$, we need to consider all possible regular trees with three inputs. There are three such trees:

The tree in the middle contributes one point to the space $W(P)(3)$ while each other tree contributes a copy of the interval $[0, 1]$ (both have only one inner edge). There is one identification to be made when the length of the inner edge is zero, in which case the corresponding tree is identified with the tree in the middle. The space $W(P)(3)$ is then the disjoint union of two copies of $[0, 1]$ where we identify the ends named 0 to a single point. What we get is that $W(P)(3) = [-1, 1]$.

It is convenient to keep track of the trees corresponding to each point in the interval $[-1, 1]$. The point $0$ corresponds to the middle tree in the picture, while a point $t \in (0, 1]$ (resp. $t \in [-1, 0)$) corresponds to the tree on the right (resp. left) where the length of the inner edge is $t$.

The case $n = 4$ is a bit more involved. There are eleven regular trees with four inputs, five of them with one internal edge and five of them with two internal edges. Here is the complete list of them:

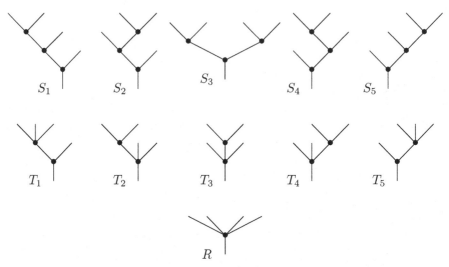

Each of the trees $S_i$ contributes a copy of the square $[0, 1] \times [0, 1]$ and each of the trees $T_i$ contributes a copy of the interval $[0, 1]$. There are several identifications when the lengths of the internal edges are zero. The space $W(P)(4)$ consists of five copies of $[0, 1] \times [0, 1]$ glued together by means of these identifications. Thus, $W(P)(4)$ is a pentagon, that we can picture as follows:

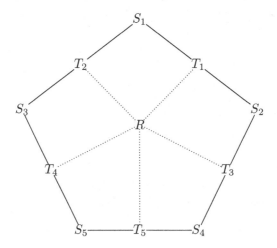

Each point in the pentagon corresponds to an element of $W(P)(4)$. The center corresponds to the tree $R$, the vertices correspond to the trees $S_i$, and the middle points of the edges correspond to the trees $T_i$ when the length of the internal edge is 1. Moving from the boundary towards the center shrinks the length of the inner edges of the trees from 1 to 0.

One can compute in this way $W(P)(n)$ for every $n$. In fact, $W(P)(n)$ is a subdivision of the $n$-th Stasheff polytope for every $n$.

*Example* 6.1.4. Let *Top* be the category of compactly generated topological spaces with the interval given by the unit interval $[0, 1]$. Let $\mathcal{C}$ be a small category considered as a discrete topological category, i.e., $\mathcal{C}(A, B)$ is viewed as a discrete topological space for every $A$ and $B$ in $\mathcal{C}$. Thus, we can view $\mathcal{C}$ as a coloured operad in *Top*, where the colours are the objects of $\mathcal{C}$ and where all operations are unary.

Then $W(\mathcal{C})$ will be again a topological operad with only unary operations, i.e., a topological category. The objects (colours) of $W(\mathcal{C})$ are the same as those of $\mathcal{C}$. The morphisms $W(\mathcal{C})(A, B)$ are represented by sequences of morphisms in $\mathcal{C}$

$$A = C_0 \xrightarrow{f_1} \overset{t_1}{C_1} \xrightarrow{f_2} \cdots \xrightarrow{f_{n-1}} \overset{t_{n-1}}{C_{n-1}} \xrightarrow{f_n} C_n = B$$

with 'waiting times' $t_i \in [0, 1]$ for $1 \leq i \leq n - 1$. If $t_i = 0$, such a sequence is identified with

$$A = C_0 \xrightarrow{f_1} \overset{t_1}{C_1} \xrightarrow{f_2} \cdots \longrightarrow \overset{t_{i-1}}{C_{i-1}} \xrightarrow{f_{i+1}f_i} \overset{t_{i+1}}{C_{i+1}} \longrightarrow \cdots \xrightarrow{f_{n-1}} \overset{t_{n-1}}{C_{n-1}} \xrightarrow{f_n} C_n = B.$$

If $f_i = \mathrm{id}$, then the sequence is identified with

$$A = C_0 \xrightarrow{f_1} \overset{t_1}{C_1} \xrightarrow{f_2} \cdots \longrightarrow \overset{s}{C_{i-1}} \xrightarrow{f_{i+1}} \overset{t_{i+1}}{C_{i+1}} \longrightarrow \cdots \xrightarrow{f_{n-1}} \overset{t_{n-1}}{C_{n-1}} \xrightarrow{f_n} C_n = B,$$

where $s = \max(t_{i-1}, t_i)$.

We will study this example in the following particular case. Let $\mathcal{C} = [n]$ be the linear tree viewed as a discrete topological category. An $[n]$-algebra in *Top* consists of a sequence of spaces $X_0, \ldots, X_n$ and maps $f_{ji} \colon X_i \longrightarrow X_j$ for $i \leq j$ such that $f_{ii} = \mathrm{id}$ and

$$f_{kj} \circ f_{ji} = f_{ki} \tag{6.1}$$

if $i \leq j \leq k$.

The topological category $W([n])$ has objects $0, 1, \ldots, n$, and a morphism $i \longrightarrow j$ in $W([n])$ is a sequence of 'times' $t_{i+1}, \ldots, t_{j-1}$ where $t_k \in [0, 1]$. In other words, $W([n])(i, j)$ is the cube $[0, 1]^{j-i-1}$ if $i + 1 \leq j$; a point if $i = j$; and the empty set if $i > j$. Composition on $W([n])$ is given by juxtaposing two such sequences putting an extra time 1 in the middle, i.e.,

$$(t_{i+1}, \ldots, t_{j-1}) \colon i \longrightarrow j \quad \text{and} \quad (t_{j+1}, \ldots, t_{k-1}) \colon j \longrightarrow k$$

compose into

$$(t_{i+1}, \ldots, t_{j-1}, 1, t_{j+1}, \ldots, t_{k-1}) \colon i \longrightarrow k.$$

A $W([n])$-algebra is then a sequence of spaces $X_0, \ldots, X_n$ and maps $f_{ji}$ as before, but for which condition (6.1) holds only up to specified coherent higher homotopies.

*Remark* 6.1.5. For any non-symmetric $C$-coloured operad $P$ in *Top*, consider all planar trees $T$ with input edges $c_1, \ldots, c_n$ and output edge $c$, such that each vertex of $T$ with input edges $b_1, \ldots, b_m$ and output edge $b$ is labelled by an element of $P(b_1, \ldots, b_m; b)$. Then

$$A(c_1, \ldots, c_n; c) \cong \coprod_T H^T,$$

where the coproduct is taken over all such trees $T$ and $H^T = H^{\times k}$, where $H = [0, 1]$ and $k$ is the number of inner edges of $T$.

The remaining identifications to construct $W(P)(c_1, \ldots, c_n; c)$ are completely determined by the combinatorics of the trees $T$. This observation is the key to generalizing the $W$-construction to coloured operads in other monoidal categories.

## 6.2   The generalized $W$-construction

In this section, we generalize the $W$-construction to coloured operads in monoidal categories. For this, one needs a suitable replacement of the unit interval $[0, 1]$ used above to give lengths to the inner edges of the trees.

**Definition 6.2.1.** Let $\mathcal{E}$ be a monoidal category with tensor product $\otimes$ and unit $I$. An *interval* in $\mathcal{E}$ is an object $H$ equipped with two 'points', i.e., maps $0, 1 \colon I \rightrightarrows H$, an augmentation $\varepsilon \colon H \longrightarrow I$ satisfying $\varepsilon \circ 0 = \mathrm{id} = \varepsilon \circ 1$, and a binary operation (playing the role of the maximum)

$$\vee \colon H \otimes H \longrightarrow H$$

which is associative, and for which 0 is unital and 1 is absorbing, i.e.,

$$0 \vee x = x = x \vee 0 \quad \text{and} \quad 1 \vee x = 1 = x \vee 1$$

for any $x \colon I \longrightarrow H$.

*Example* 6.2.2. The unit interval $[0, 1]$ is an interval in *Top*. One can choose as $\vee$ the maximum operation or the 'reversed' multiplication, i.e., $s \vee t$ is defined by the identity $(1 - s \vee t) = (1 - s)(1 - t)$.

The groupoid $J = (0 \overset{\sim}{\longleftrightarrow} 1)$ is an interval in *Cat*. Another possible interval for *Cat* is the two-object category $I = (0 \longrightarrow 1)$.

For any interval $H$ and any coloured operad $P$ in $\mathcal{E}$, there is a new coloured operad $W_H(P)$ (on the same colours as $P$) together with a natural map of operads $W_H(P) \longrightarrow P$. The operad $W_H(P)$ is constructed as $W(P)$ in the case of topological spaces, now glueing objects of the form $H^{\otimes k}$ instead of cubes $[0,1]^k$ (see Remark 6.1.5). The functor $W_H$ is called the *W*-construction in $\mathcal{E}$ associated to the interval $H$. As was pointed out in the case of topological operads, the *W*-construction can be defined in a similar way for symmetric coloured operads in $\mathcal{E}$ (cf. Remark 6.1.2). We refer the reader to [BM06] and [BM07] for more details on the generalized *W*-construction.

*Example* 6.2.3. Let *Cat* be the category of (small) categories with the groupoid interval $J = (0 \xleftrightarrow{\sim} 1)$. Let $I$ denote the unit for the cartesian product of categories, i.e., $I$ is the category with one object and one (identity) morphism. Consider a non-symmetric operad $P$ in *Cat* with $P(n) = I$ corresponding to one $n$-ary operation for every $n \geq 1$, and $P(0) = \emptyset$.

The operad $W_J(P)$ is a one-colour operad and, as in Example 6.1.3, the term $W_J(P)(n)$ is described by using regular trees (i.e., trees with no unary vertices) with $n$ leaves. If $n = 1, 2$, then $W_J(P)(n)$ is the one-object category $I$. For $n = 3$, there are three regular trees, two with one inner edge (each contributing a copy of $J$) and one without inner edges (contributing a copy of $I$). In the *W*-construction, we identify the object named 0 in every copy of $J$ with the unique object of $I$. Hence, $W_J(P)(3)$ is a category with three objects and a unique isomorphism between any two objects. We can picture it as

$$1 \xleftrightarrow{\sim} 0 \xleftrightarrow{\sim} 1.$$

Similarly, in the case $n = 4$, there are eleven regular trees with four inputs (see Example 6.1.3), five of them with two internal edges and five of them with one internal edge. Each of the trees with two internal edges contributes a copy of $J \times J$ to $W_J(P)(4)$. Using the identifications given by the *W*-construction, one can show that $W_J(P)(4)$ consists of a category with eleven objects and a unique isomorphism between any two objects, and we can picture it as

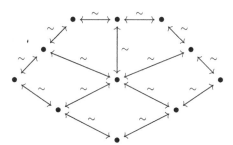

# 6.3 The homotopy coherent nerve

In this section, we use the generalized *W*-construction to define, for every coloured operad $P$ in a symmetric monoidal category, a dendroidal set $hcN_d(P)$ called the *homotopy coherent nerve* of $P$. This dendroidal set is similar to the one obtained via the dendroidal nerve construction, but with homotopies built into it.

Let $\mathcal{E}$ be a symmetric monoidal category with an interval $H$. For each tree $T$ in $\Omega$ we can consider the operad $\Omega(T)$ as a discrete operad in $\mathcal{E}$. This is done by applying to the operad $\Omega(T)$ in *Sets* the strong monoidal functor *Sets* $\longrightarrow \mathcal{E}$ sending any set $X$ to $\coprod_{x \in X} I$, where $I$ is the unit of $\mathcal{E}$. Then we have a functor

$$\Omega \longrightarrow Oper(\mathcal{E})$$

that assigns to each tree $T$ the operad $W_H(\Omega(T))$. By Kan extension, this functor induces a pair of adjoint functors

$$hc\tau_d : dSets \rightleftarrows Oper(\mathcal{E}) : hcN_d. \tag{6.2}$$

**Definition 6.3.1.** The functor $hcN_d \colon Oper(\mathcal{E}) \longrightarrow dSets$ is called the *homotopy coherent nerve* functor.

More explicitly, for every tree $T$ in $\Omega$ and any coloured operad $P$ in $\mathcal{E}$, the homotopy coherent nerve is given by

$$hcN_d(P)_T = Oper(W_H(\Omega(T)), P).$$

To have a better understanding of the functor $hcN_d$ it will be useful to have a description of the operad $W_H(\Omega(T))$. We do it in the case $\mathcal{E} = Top$, although a similar description applies to any monoidal category $\mathcal{E}$ with an interval.

*Example* 6.3.2. Let $\mathcal{E} = Top$ be the category of compactly generated topological spaces with the interval $H = [0,1]$. Let $T$ be any tree in $\Omega$. The colours of the operad $W(\Omega(T))$ are the same as those of $\Omega(T)$, i.e., the edges of $T$. If $\sigma = (e_1, \ldots, e_n, e)$ are edges of $T$ such that there is a subtree $T_\sigma$ of $T$ with $e_1, \ldots, e_n$ as input edges and $e$ as output edge, then

$$W(\Omega(T))(e_1, \ldots, e_n; e) = H^{\#\mathrm{inn}(T_\sigma)}$$

where $\mathrm{inn}(T_\sigma)$ is the set of inner edges of $T_\sigma$. In fact, $W(\Omega(T))(e_1, \ldots, e_n; e)$ is a $\#\mathrm{inn}(T_\sigma)$-dimensional cube representing the space of assignments of lengths $t_i \in [0,1]$ to the internal edges of $T_\sigma$. It is the one-point space if $\mathrm{inn}(T_\sigma)$ is the empty set.

If there is no subtree in $T$ with $e_1, \ldots, e_n$ as input edges and $e$ as output edge, then

$$W(\Omega(T))(e_1, \ldots, e_n; e) = \emptyset.$$

The composition operations in $W(\Omega(T))$ are given in terms of the $\circ_i$ operations as follows. If $\sigma = (e_1, \ldots, e_n; e)$ and $\sigma' = (d_1, \ldots, d_m; e_i)$ represent two subtrees of $T$, then the composition map

$$W(\Omega(T))(e_1, \ldots, e_n; e) \times W(\Omega(T))(d_1, \ldots, d_m; e_i)$$

$$\downarrow \circ_i$$

$$W(\Omega(T))(e_1, \ldots, e_{i-1}, d_1, \ldots, d_m, e_{i+1}, \ldots, e_n; e)$$

is defined by grafting the trees $T_\sigma$ and $T_{\sigma'}$ along the edge $e_i$ to form another subtree $T_{\sigma \circ_i \sigma'}$. This new tree $T_{\sigma \circ_i \sigma'}$ has as internal edges the ones of $T_\sigma$ and the ones of $T_{\sigma'}$ plus a new one $e_i$ which is assigned length 1.

The left adjoint $hc\tau_d$ is closely related to the $W$-construction. Let $P$ be a coloured operad in *Sets* and let $P_\mathcal{E}$ be the corresponding operad in $\mathcal{E}$ obtained via the functor *Sets* $\longrightarrow \mathcal{E}$. Then we have the following proposition:

**Proposition 6.3.3.** *Let $\mathcal{E}$ be a symmetric monoidal category with an interval $H$ and let $P$ be any coloured operad in Sets. Then there is a natural isomorphism of operads*

$$hc\tau_d(N_d(P)) \cong W_H(P_\mathcal{E}).$$

*Proof.* The dendroidal set $N_d(P)$ is a colimit of representables $\Omega[T]$ over all morphisms $\Omega[T] \longrightarrow N_d(P)$. Since, by adjunction,

$$dSets(\Omega[T], N_d(P)) \cong Oper(\tau_d(\Omega[T]), P) \cong Oper(\Omega(T), P),$$

it follows that

$$N_d(P) = \varinjlim_{\Omega(T) \to P} \Omega[T].$$

Using the fact that $hc\tau_d$ preserves colimits and that $hc\tau_d(\Omega[T]) \cong W_H(\Omega(T))$, we have that

$$hc\tau_d(N_d(P)) = \varinjlim_{\Omega(T) \to P} W_H(\Omega(T)).$$

The required isomorphism follows now by direct inspection of the explicit construction of $W_H(P_\mathcal{E})$ given in [BM06]. $\square$

An immediate consequence of this result is that there is a natural bijection

$$dSets(N_d(P), hcN_d(Q)) \cong Oper(W_H(P_\mathcal{E}), Q).$$

More generally, we have the following theorem:

**Theorem 6.3.4.** *Let $\mathcal{E}$ be a symmetric monoidal category with an interval $H$. Then there is a natural isomorphism*

$$\mathrm{Hom}_{dSets}(N_d(P), hcN_d(Q))_T \cong Oper(W_H(P_\mathcal{E} \otimes_{BV} \Omega(T)), Q)$$

*for every tree $T$ in $\Omega$ and coloured operads $P$ and $Q$ in sets.*

*Proof.* By the definition of the internal hom in dendroidal sets, we have that

$$\mathrm{Hom}_{dSets}(N_d(P), hcN_d(Q))_T = dSets(N_d(P) \otimes \Omega[T], hcN_d(Q)).$$

The required natural isomorphism follows now from the adjunction (6.2) and Proposition 6.3.3, using the fact that

$$N_d(P) \otimes \Omega[T] \cong N_d(P \otimes_{BV} \Omega(T))$$

for every operad $P$ and every tree $T$ in $\Omega$. $\square$

# Lecture 7

# Inner Kan complexes and normal dendroidal sets

## 7.1 Inner Kan complexes

In this section, we introduce the notion of inner Kan complexes in the category of dendroidal sets. We begin by recalling the definition of an inner Kan complex in the category of simplicial sets. A horn $\Lambda^k[n]$ in simplicial sets is called an *inner horn* if $0 < k < n$.

**Definition 7.1.1.** A simplicial set $X$ satisfies the *restricted Kan condition* if every inner horn $f\colon \Lambda^k[n] \longrightarrow X$ has a filler, i.e., there exists a map $g\colon \Delta[n] \longrightarrow X$ such that $f = g \circ j$, where $j\colon \Lambda^k[n]\rightarrowtail\Delta[n]$ denotes the inclusion

$$\Lambda^k[n] \xrightarrow{\ f\ } X, \qquad (7.1)$$
$$j \downarrow \quad \nearrow g$$
$$\Delta[n]$$

or, equivalently, if the induced map

$$j^*\colon sSets(\Delta[n], X) \longrightarrow sSets(\Lambda^k[n], X) \qquad (7.2)$$

is surjective for every $0 < k < n$.

    A simplicial set satisfying the restricted Kan condition is called an *inner Kan simplicial set* (a *quasi-category* in Joyal's terminology). If the filler of (7.1) is unique or, equivalently, if the map (7.2) is a bijection, then $X$ is called a *strict inner Kan simplicial set.*

I. Moerdijk and B. Toën, *Simplicial Methods for Operads and Algebraic Geometry*, Advanced Courses in Mathematics - CRM Barcelona, DOI 10.1007/978-3-0348-0052-5_7, © Springer Basel AG 2010

The definition of an inner Kan complex for dendroidal sets is similar to the one for simplicial sets, but using the inner horns defined in Lecture 3. Recall from Section 3.2 that if $T$ is any tree, $e$ is any inner edge of $T$, and $\partial_e$ is the face map in $\Omega$ contracting $e$, then the inner horn $\Lambda^e[T]$ is defined as

$$\Lambda^e[T] = \bigcup_{\beta \neq \partial_e \in \Phi_1(T)} \partial_\beta \Omega[T],$$

where $\Phi_1(T)$ is the set of all faces of $T$ and $\partial_\beta \Omega[T]$ is the $\beta$-face of $\Omega[T]$.

The inner horn $\Lambda^e[T]$ is a subobject of the boundary $\partial\Omega[T]$ and extends the notion of inner horn for simplicial sets, namely

$$i_!(\Lambda^k[n]) = \Lambda^k[L_n] \tag{7.3}$$

as a subobject of $i_!(\Delta[n]) = \Omega[L_n]$, where $L_n = i([n])$ is the linear tree with $n$ vertices and $n+1$ edges.

**Definition 7.1.2.** A dendroidal set $X$ is an *inner Kan complex* if, for every tree $T$, every inner horn $f \colon \Lambda^e[T] \longrightarrow X$ has a filler, i.e., there exists a map $g \colon \Omega[T] \longrightarrow X$ such that $f = g \circ j$, where $j \colon \Lambda^e[T] \longrightarrow \Omega[T]$ denotes the inclusion

$$
\begin{array}{ccc}
\Lambda^e[T] & \xrightarrow{\ f\ } & X, \\
{\scriptstyle j}\downarrow & \nearrow & \\
\Omega[T] & {\scriptstyle g} &
\end{array}
\tag{7.4}
$$

or, equivalently, if the induced map

$$j^* \colon dSets(\Omega[T], X) \longrightarrow dSets(\Lambda^e[T], X) \tag{7.5}$$

is surjective for every tree $T$ and every inner edge $e$ of $T$.

If the filler of (7.4) is unique or, equivalently, if the map (7.5) is a bijection, then $X$ is called a *strict inner Kan complex*.

A map $f \colon X \longrightarrow Y$ of dendroidal sets is an *inner Kan fibration* if it has the left lifting property with respect to any inner horn inclusion $\Lambda^e[T] \longrightarrow \Omega[T]$, for every tree $T$ and every inner edge $e$ of $T$. Thus, a dendroidal set $X$ is an inner Kan complex if the map $X \longrightarrow 1$ is an inner Kan fibration, where 1 denotes the terminal object of the category $dSets$.

**Proposition 7.1.3.** *Let $K$ be any simplicial set and let $X$ be any dendroidal set. Then:*

(i) *The dendroidal set $i_!(K)$ is an inner Kan complex if and only if $K$ is an inner Kan simplicial set.*

(ii) *If $X$ is an inner Kan complex, then the simplicial set $i^*(X)$ is an inner Kan simplicial set.*

*Proof.* The results follow immediately from the fact that $i_!$ is fully faithful, the adjunction between $i_!$ and $i^*$, and (7.3). $\qquad\square$

A source of strict inner Kan complexes is given by dendroidal nerves of operads (see Example 3.1.4).

**Proposition 7.1.4.** *Let $P$ be any coloured operad in Sets. Then $N_d(P)$ is a strict inner Kan complex.*

*Proof.* Any dendrex $x \in N_d(P)_T$ is given by a map $x\colon \Omega[T] \longrightarrow N_d(P)$, which corresponds by the adjunction (3.1) to a map of operads $\Omega(T) \longrightarrow P$. If we choose a planar representative for $T$, then $\Omega(T)$ is a free operad generated by the operations corresponding to the vertices of the (planar representative of the) tree $T$. It follows that $x$ is equivalent to a labeling of the (planar representative of the) tree $T$ as follows. The edges of $T$ are labeled by the colours of $P$ and the vertices are labeled by operations in $P$, where the inputs of such an operation are given by the labels of the incoming edges to the vertex, and the output is the label of the outgoing edge from the vertex. Any inner horn $\Lambda^e[T] \longrightarrow N_d(P)$ completely determines such a labeling of the tree $T$ and thus determines a unique extension $\Omega(T) \longrightarrow P$. $\qquad\square$

*Remark* 7.1.5. The converse of this result is also true, as we will prove in Section 7.3.

**Proposition 7.1.6.** *Any strict inner Kan complex $X$ is 2-coskeletal.*

*Proof.* Let $X$ be a strict inner Kan complex and let $A$ be any dendroidal set. Suppose that a map $f\colon \mathrm{Sk}_2 A \longrightarrow \mathrm{Sk}_2 X$ is given. We first show that this map $f$ can be extended to a dendroidal map $\hat{f}\colon A \longrightarrow X$. Assume that $f$ was extended to a map $f_k\colon \mathrm{Sk}_k A \longrightarrow \mathrm{Sk}_k X$ for $k \geq 2$. Let $a \in \mathrm{Sk}_{k+1}(A)$ be a non-degenerate dendrex and suppose that $a \notin \mathrm{Sk}_k(A)$. Hence, $a \in A_T$ and $T$ has exactly $k+1$ vertices. Now, we choose an inner horn $\Lambda^e[T]$, which always exists since $k \geq 2$. The set $\{\beta^* a\}_{\beta \neq \partial_e}$ where $\beta$ runs over all faces of $T$ defines a horn $\Lambda^e[T] \longrightarrow A$. Since this horn factors through the $k$-skeleton of $A$, we obtain, by composition with $f_k$, a horn $\Lambda^e[T] \longrightarrow X$ in $X$ given by $\{f(\beta^* a)\}_{\beta \neq \alpha}$. If $f_{k+1}(a) \in X_T$ denotes the unique filler of that horn, then we have that $\beta^* f_{k+1}(a) = f(\beta^* a)$ for each $\beta \neq \partial_e$.

To obtain the same property for $\partial_e$, observe that the dendrices $f(\partial_e^* a)$ and $\partial_e^* f_{k+1}(a)$ both have the same boundary and that they are both of shape $S$, where $S$ has $k$ vertices. Since $k \geq 2$, we have that $S$ has an inner face, but then it follows that both $f(\partial_e^* a)$ and $\partial_e^* f_{k+1}(a)$ are fillers for the same inner horn in $X$ and hence equal. If we repeat this process for all dendrices in $\mathrm{Sk}_{k+1}(A)$, we get that $f_k$ can be extended to $f_{k+1}\colon \mathrm{Sk}_{k+1}(A) \longrightarrow \mathrm{Sk}_{k+1}(X)$. This holds for all $k \geq 2$, which implies that $f$ can be extended to $\hat{f}\colon A \longrightarrow X$.

In order to prove the uniqueness of $\hat{f}$, assume that $g$ is another extension of $f$ and that it has been shown that $\hat{f}$ and $g$ agree on all dendrices of shape $T$ where $T$ has at most $k$ vertices. Let $a \in X_S$ be a dendrex of shape $S$, where $S$ has $k+1$ vertices. Then the dendrices $\hat{f}(a)$ and $g(a)$ are dendrices in $X$ that have the same boundary. Since $k \geq 2$, it follows that these dendrices are both fillers for the same inner horn, hence they are the same and thus $\hat{f} = g$. □

## 7.2  Inner anodyne extensions

In this section, we develop the notion of inner anodyne extensions for dendroidal sets.

**Definition 7.2.1.** Let $\mathcal{M}$ be a class of monomorphisms in *dSets*. We say that $\mathcal{M}$ is *saturated* if it contains all the isomorphisms and it is closed under pushouts, retracts, arbitrary coproducts, and colimits of sequences (indexed by ordinals).

Given an arbitrary class of monomorphisms $\mathcal{M}$, the *saturated class generated by* $\mathcal{M}$ is the smallest saturated class that contains $\mathcal{M}$, i.e., the intersection of all the saturated classes containing $\mathcal{M}$.

**Definition 7.2.2.** The class of *inner anodyne extensions* in *dSets* is the saturated class generated by the set of inner horn inclusions. Thus, it is also the class of maps having the left lifting property with respect to the inner Kan fibrations.

The surjectivity property for inner Kan complexes extends to inner anodyne extensions, namely if $u \colon U \longrightarrow V$ is an inner anodyne extension, then the map

$$u^* \colon dSets(V, K) \longrightarrow dSets(U, K)$$

is surjective for any inner Kan complex $K$. Similarly, $u^*$ is a bijection for any strict inner Kan complex $K$.

Given any tree $T$, let $I(T)$ be the subset of the edges of $T$ consisting of only the inner edges. For any nonempty subset $E \subset I(T)$, we denote by $\Lambda^E[T]$ the union of all the faces of $\Omega[T]$ except those obtained by contracting an edge from $E$, i.e.,

$$\Lambda^E[T] = \bigcup_{\alpha \in \Phi_1(T) \setminus \partial_E} \partial_\alpha \Omega[T],$$

where $\partial_E = \{ \partial_e \mid e \in E \}$. Observe that, if $E = \{e\}$, then $\Lambda^E[T] = \Lambda^e[T]$.

**Lemma 7.2.3.** *For any nonempty $E \subseteq I(T)$, the inclusion $\Lambda^E[T] \rightarrowtail \Omega[T]$ is inner anodyne.*

*Proof.* We will proceed by induction on the number $n$ of elements of $E$. If $n = 1$, then $\Lambda^E[T] \rightarrowtail \Omega[T]$ is an inner horn inclusion, thus inner anodyne.

Assume that the result holds for $n < k$ and suppose that $E$ has $k$ elements. Let $e$ be any element of $E$ and let $F = E \setminus \{e\}$. Then the map $\Lambda^E[T] \longrightarrow \Omega[T]$ factors as

$$\Lambda^E[T] \longrightarrow \Lambda^F[T]$$
$$\searrow \quad \downarrow$$
$$\Omega[T],$$

where the vertical map in the diagram is inner anodyne by the induction hypothesis, since $F$ has $k-1$ elements. We can express the horizontal map as the following pushout:

$$\begin{array}{ccc} \Lambda^F[T/e] & \longrightarrow & \Lambda^E[T] \\ \downarrow & & \downarrow \\ \Omega[T/e] & \longrightarrow & \Lambda^F[T]. \end{array}$$

Now, the map on the left is inner anodyne (again by the induction hypothesis), hence so is the map on the right, since inner anodyne extensions are closed under pushouts. Therefore $\Lambda^E[T] \longrightarrow \Omega[T]$ is a composition of two inner anodyne extensions and thus it is inner anodyne too. $\qquad\square$

The above lemma implies that $\Lambda^I[T] \longrightarrow \Omega[T]$ is inner anodyne, where $\Lambda^I[T]$ is an abbreviation for $\Lambda^{I(T)}[T]$.

We now consider how dendrices in an inner Kan complex can be grafted. Recall that for any two trees $T$ and $S$, and $l$ a leaf of $T$, we denote by $T \circ_l S$ the tree obtained by grafting $S$ onto $T$ by identifying $l$ with the root of $S$. Both $S$ and $T$ naturally embed as subfaces of $T \circ_l S$, which induces the obvious inclusions $\Omega[S] \rightarrowtail \Omega[T \circ_l S]$ and $\Omega[T] \rightarrowtail \Omega[T \circ_l S]$, the pushout of which we denote by $\Omega[T] \cup_l \Omega[S] \longrightarrow \Omega[T \circ_l S]$.

**Lemma 7.2.4.** *For any two trees $T$ and $S$ and any leaf $l$ of $T$, the inclusion*

$$\Omega[T] \cup_l \Omega[S] \longrightarrow \Omega[T \circ_l S]$$

*is an inner anodyne extension.*

*Proof.* We may assume that $T \neq | \neq S$, otherwise the result is obvious. We proceed by induction on the sum $n$ of the number of vertices of $T$ and $S$. Let $R = T \circ_l S$. If $n = 2$, then $\Omega[T] \cup_l \Omega[S] \longrightarrow \Omega[T \circ_l S]$ is a horn inclusion, and thus inner anodyne.

Assume that the result holds for $2 \leq n < k$ and that the sum of the number of vertices of $T$ and $S$ is $k$. Recall that $\Lambda^I[R]$ is the union of all the outer faces of $\Omega[R]$. Observe that the map $\Omega[T] \cup_l \Omega[S] \longrightarrow \Omega[R]$ factors as

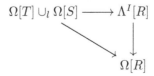

and that the vertical map is inner anodyne by Lemma 7.2.3. We will show that
the horizontal map is inner anodyne by expressing it as a pushout of an inner
anodyne extension.

For the purpose of this proof, let us say that an *external cluster* of a tree is a
vertex $v$ with the property that one of the edges adjacent to it is inner while all the
rest are outer. Let $\mathrm{Cl}(T)$ (resp. $\mathrm{Cl}(S)$) denote the set of all external clusters of $T$
(resp. of $S$) which do not contain $l$ (resp. the root of $S$). For each vertex $v \in \mathrm{Cl}(T)$,
the face $\partial_v$ of $\Omega[R]$ corresponding to $v$ is isomorphic to $\Omega[(T/v) \circ_l S]$ and the map
$\Omega[T/v] \cup_l \Omega[S] \longrightarrow \Omega[(T/v) \circ_l S]$ is inner anodyne by the induction hypothesis. In
a similar way, for every $w \in \mathrm{Cl}(S)$ the face $\partial_w$ of $\Omega[R]$ that corresponds to $w$ is
isomorphic to $\Omega[T \circ_l (S/w)]$ and the map $\Omega[T] \cup_l \Omega[S/w] \longrightarrow \Omega[T \circ_l (S/w)]$ is inner
anodyne, again by the induction hypothesis. The following diagram is a pushout:

$$
\left( \coprod_{c\in\mathrm{Cl}(T)} (\Omega[T/c] \cup_l \Omega[S]) \right) \amalg \left( \coprod_{c\in\mathrm{Cl}(T)} (\Omega[T] \cup_l \Omega[S/c]) \right) \longrightarrow \Omega[T] \cup_l \Omega[S]
$$

$$
\downarrow \qquad\qquad\qquad\qquad\qquad\qquad\qquad\qquad\qquad\qquad\qquad \downarrow
$$

$$
\left( \coprod_{c\in\mathrm{Cl}(T)} (\Omega[(T/c) \circ_l S]) \right) \amalg \left( \coprod_{c\in\mathrm{Cl}(T)} (\Omega[T \circ_l (S/c)]) \right) \longrightarrow \Lambda^I[R],
$$

where the left vertical map is the coproduct of all the inner anodyne extensions
mentioned above, thus inner anodyne. This implies that the vertical map on the
right is inner anodyne too.                                        □

## 7.3   Homotopy in an inner Kan complex

In this section, we study a notion of homotopy inside dendroidal sets. Two dendrices are said to be homotopic if one is a composition of the other with a degenerate dendrex. We will show that this homotopy theory within a dendroidal set
is well behaved if the dendroidal set is an inner Kan complex. In that case, the
resulting homotopy relation is an equivalence relation. From this it follows that to
every inner Kan complex $X$ we can associate an operad $\mathrm{Ho}(X)$, which we call the
*homotopy operad* associated to $X$. The aim of this section is to prove a converse
of Proposition 7.1.4, namely

**Theorem 7.3.1.** *A dendroidal set $X$ is a strict inner Kan complex if and only if
$X$ is the dendroidal nerve of an operad.*

For each $n \geq 0$, let $C_n$ be the $n$-th corolla:

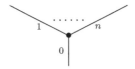

For each $0 \leq i \leq n$, we denote by $i \colon \eta \longrightarrow C_n$ the outer face map in $\Omega$ that sends the unique edge of $\eta$ to the edge $i$ in $C_n$. An element $f \in X_{C_n}$ will be denoted by

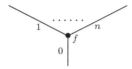

**Definition 7.3.2.** Let $X$ be an inner Kan complex and let $f, g \in X_{C_n}$ and $n \geq 0$. For $1 \leq i \leq n$, we say that $f$ *is homotopic to* $g$ *along the edge* $i$, and denote it by $f \sim_i g$, if there is a dendrex $H$ of shape

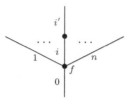

whose three faces are:

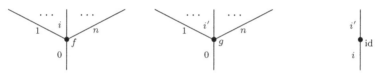

where the 'id' in the third tree is a degeneracy of $i$. In the same way, we will say that $f$ is homotopic to $g$ along the edge $0$, and denote it by $f \sim_0 g$, if there is a dendrex of shape

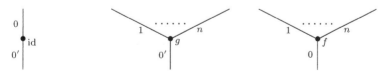

whose three faces are:

When $f \sim_i g$ for some $0 \leq i \leq n$, we will refer to the corresponding $H$ as a *homotopy from* $f$ *to* $g$ *along* $i$.

*Remark* 7.3.3. Notice that, in a strict inner Kan complex $X$, the homotopy relation just defined is the identity relation.

**Proposition 7.3.4.** *Let $X$ be an inner Kan complex. The relations $\sim_i$ on the set $X_{C_n}$ are equivalence relations for each $0 \leq i \leq n$, and these equivalence relations all coincide.*

*Proof.* For a detailed proof, see [MW09, Proposition 6.3 and Lemma 6.4]. $\square$

Due to this proposition, we will use the notation $f \sim g$ instead of $f \sim_i g$. Given an inner Kan complex $X$ and $x_1, \ldots, x_n, x \in X_\eta$, we denote by

$$X(x_1, \ldots, x_n; x) \subseteq X_{C_n}$$

the set of all dendrices $f$ such that $0^*(f) = x$ and $i^*(f) = x_i$ for $1 \leq i \leq n$. We can define a coloured collection $\mathrm{Ho}(X)$ as follows. The set of colours is the set $X_\eta$, and given objects $x_1, \ldots, x_n, x \in X_\eta$, we define

$$\mathrm{Ho}(x_1, \ldots, x_n; x) = X(x_1, \ldots, x_n; x_0)/\sim$$

where $\sim$ is the equivalence relation on $X_{C_n}$ given by Proposition 7.3.4. In order to put an operad structure on the collection $\mathrm{Ho}(X)$, we need to define the composition operations $\circ_i$.

**Definition 7.3.5.** Let $X$ be an inner Kan complex and let $f \in X_{C_n}$ and $g \in X_{C_m}$ be two dendrices in $X$. A dendrex $h \in X_{C_{n+m-1}}$ is a $\circ_i$-*composition* of $f$ and $g$, denoted by $h \sim f \circ_i g$, if there is a dendrex $\lambda$ in $X$,

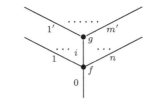

with inner face

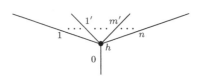

The dendrex $\lambda$ is called a *witness* for the composition.

*Remark 7.3.6.* Notice that, for $1 \leq i \leq n$, we have by definition that $H: f \sim_i g$ if and only if $H$ is a witness for the composition $g \sim f \circ_i \mathrm{id}$. Similarly, for $i = 0$ we have that $H: f \sim_0 g$ if and only if $H$ is a witness for the composition $g \sim \mathrm{id} \circ f$.

If $X$ is an inner Kan complex and $f \sim f'$ and $g \sim g'$, then, if $h \sim f \circ_i g$ and $h' \sim f' \circ_i g'$, we have that $h \sim h'$ (see [MW09, Lemma 6.9]). Hence composition is well defined on homotopy classes.

**Proposition 7.3.7.** *There is a unique structure of a symmetric coloured operad on* $\mathrm{Ho}(X)$ *for which the canonical map of collections* $\mathrm{Sk}_1(X) \longrightarrow \mathrm{Ho}(X)$ *extends to a map of dendroidal sets* $X \longrightarrow N_d(\mathrm{Ho}(X))$. *The latter map is an isomorphism whenever* $X$ *is a strict inner Kan complex.*

*Proof.* Given $[f] \in \mathrm{Ho}(X)(x_1, \ldots, x_n; x)$ and $[g] \in \mathrm{Ho}(X)(y_1, \ldots, y_m; x_i)$, the assignment

$$[f] \circ_i [g] = [f \circ_i g]$$

is well defined by Remark 7.3.6. This gives the $\circ_i$ operations in the coloured operad $\mathrm{Ho}(X)$. The actions of the symmetric group $\Sigma_n$ are defined in the following way. For any element $\sigma \in \Sigma_n$, let $\sigma \colon C_n \longrightarrow C_n$ be the induced map in $\Omega$ that permutes the edges of the $n$-th corolla. The map $\sigma^* \colon X_{C_n} \longrightarrow X_{C_n}$ restricts to a function

$$\sigma^* \colon X(x_1, \ldots, x_n; x) \longrightarrow X(x_{\sigma(1)}, \ldots, x_{\sigma(n)}; x)$$

that respects the homotopy relation. Hence, we get a map

$$\sigma^* \colon \mathrm{Ho}(X)(x_1, \ldots, x_n; x) \longrightarrow \mathrm{Ho}(X)(x_{\sigma(1)}, \ldots, x_{\sigma(n)}; x).$$

It is straightforward that these structure maps provide the coloured collection $\mathrm{Ho}(X)$ with an operad structure.

To prove that the quotient map $q \colon \mathrm{Sk}_1(X) \longrightarrow \mathrm{Ho}(X)$ extends to a map $q \colon X \longrightarrow N_d(\mathrm{Ho}(X))$ of dendroidal sets, it is enough to give its values for dendrices $x \in X_T$ where $T$ is a tree with two vertices, since $N_d(\mathrm{Ho}(X))$ is 2-coskeletal by Proposition 7.1.6. So, let $T$ be a tree with two vertices and $e$ be the inner edge of this tree. Then the map

$$\Lambda^e[T] \rightarrowtail \Omega[T] \xrightarrow{x} X$$

factors through $\mathrm{Sk}_1(X)$, so its composition $\Lambda^e[T] \longrightarrow N_d(\mathrm{Ho}(X))$ with $q$ has a unique extension by Proposition 7.1.4. We take this extension to be $q(x) \colon \Omega[T] \longrightarrow N_d(\mathrm{Ho}(X))$ and this defines the map $q \colon \mathrm{Sk}_2(X) \longrightarrow \mathrm{Sk}_2(N_d(\mathrm{Ho}(X)))$, and hence all of $q \colon X \longrightarrow N_d(\mathrm{Ho}(X))$ by 2-coskeletality.

If $X$ is itself a strict inner Kan complex, then the homotopy relation is the identity relation, so $\mathrm{Sk}_1(X) \longrightarrow \mathrm{Ho}(X)$ is the identity map. Since $X$ and $N_d(\mathrm{Ho}(X))$ are strict inner Kan complexes, it follows that $q \colon X \longrightarrow N_d(\mathrm{Ho}(X))$ is an isomorphism. $\square$

The following result together with Proposition 7.3.7 provide the proof of Theorem 7.3.1.

**Proposition 7.3.8.** *The natural map* $\tau_d(X) \longrightarrow \mathrm{Ho}(X)$ *is an isomorphism of operads for every inner Kan complex* $X$.

*Proof.* It is enough to prove that the map $q \colon X \longrightarrow N_d(\mathrm{Ho}(X))$ of Proposition 7.3.7 has the universal property of the unit of the adjunction. This means that, for any operad $P$ and any map $\varphi \colon X \longrightarrow N_d(P)$, there is a unique map of operads $\psi \colon \mathrm{Ho}(X) \longrightarrow P$ such that $N_d(\psi)q = \varphi$. Observe that $\varphi$ induces a map $\mathrm{Ho}(X) \longrightarrow \mathrm{Ho}(N_d(P))$ such that the diagram

$$
\begin{array}{ccc}
\mathrm{Sk}_1(X) & \xrightarrow{\ \varphi\ } & \mathrm{Sk}_1 N_d(P) \\
\downarrow{\scriptstyle q} & & \downarrow{\scriptstyle q_P} \\
\mathrm{Ho}(X) & \xrightarrow{\ \mathrm{Ho}(\varphi)\ } & \mathrm{Ho}(N_d(P))
\end{array}
$$

commutes. Now, $\mathrm{Ho}(N_d(P)) = P$ and $q_P$ is the identity, as we can see from the proof of Proposition 7.3.7. Hence, $\mathrm{Ho}(\varphi)$ defines a map $\psi \colon \mathrm{Ho}(X) \longrightarrow P$ of coloured collections. In fact, one can see that $\psi$ is a map of operads, and the uniqueness follows from the surjectivity of $q$.  □

*Proof of Theorem* 7.3.1. One direction was already proved in Proposition 7.1.4. For the other one, suppose that $X$ is a strict inner Kan complex. Then Proposition 7.3.7 and Proposition 7.3.8 imply that $X \cong N_d(\tau_d(X))$.  □

## 7.4   Homotopy coherent nerves are inner Kan

Throughout this section, $\mathcal{E}$ will denote a monoidal model category with cofibrant unit $I$. We will also assume that $\mathcal{E}$ is equipped with an *interval* in the sense of [BM06], which we denote by $H$. Recall from Definition 6.2.1 that such an interval is given by an object $H$ of $\mathcal{E}$ together with maps

$$
I \underset{1}{\overset{0}{\rightrightarrows}} H \xrightarrow{\ \epsilon\ } I \qquad \text{and} \qquad H \otimes H \xrightarrow{\ \vee\ } H
$$

satisfying certain conditions. This means in particular that $H$ is an interval in Quillen's sense ([Qui67]), so 0 and 1 together define a cofibration $I \coprod I \longrightarrow H$, and $\epsilon$ is a weak equivalence. In Section 6.2 we explained how such an interval $H$ allows one to construct for each coloured operad $P$ in $\mathcal{E}$ a Boardman–Vogt resolution $W_H(P) \longrightarrow P$. Each operad in *Sets* can be viewed as an operad in $\mathcal{E}$ via the strong monoidal functor *Sets* $\longrightarrow \mathcal{E}$, as in Example 1.3.6, and hence has such a Boardman–Vogt resolution. When we apply this construction to the operads $\Omega(T)$, we obtain the *homotopy coherent dendroidal nerve* $hcN_d(P)$ of any operad $P$ in $\mathcal{E}$, defined as the dendroidal set given by

$$
hcN_d(P)_T = Oper(W_H(\Omega(T)), P).
$$

The goal of this section is to prove that the homotopy coherent nerve is an inner Kan complex.

**Theorem 7.4.1.** *Let $P$ be a $C$-coloured operad in $\mathcal{E}$ such that, for every $(n+1)$-tuple $(c_1, \ldots, c_n, c)$ of colours of $P$, the object $P(c_1, \ldots, c_n; c)$ is fibrant in $\mathcal{E}$. Then $hcN_d(P)$ is an inner Kan complex.*

As observed in Section 6.3, our construction of the dendroidal homotopy coherent nerve specializes to that of the homotopy coherent nerve of an $\mathcal{E}$-enriched category. In the case where $\mathcal{E}$ is the category of topological spaces or simplicial sets, one recovers the classical definition of Cordier and Porter ([CP86]). It follows as a particular case of Theorem 7.4.1 that the homotopy coherent nerve of an $\mathcal{E}$-enriched category with fibrant Hom objects is a quasi-category in the sense of Joyal. This was proved in [CP86] when $\mathcal{E}$ is the category of simplicial sets.

Recall from Example 6.3.2, in the case $\mathcal{E} = Top$, the description of the operads $W_H(\Omega(T))$ involved in the definition of the homotopy coherent nerve. We have a similar description for these operads in the case of a general monoidal model category $\mathcal{E}$. First of all, recall from (1.1) the symmetrization functor

$$\Sigma \colon Oper_{\Sigma}(\mathcal{E}) \longrightarrow Oper(\mathcal{E}),$$

which is left adjoint to the forgetful functor from symmetric operads to non-symmetric ones. If $T$ is a tree in $\Omega$, then any planar representative $\bar{T}$ of $T$ naturally describes a non-$\Sigma$ operad $\Omega(\bar{T})$ such that $\Omega(T) = \Sigma(\Omega(\bar{T}))$. It follows that

$$W_H(\Omega(T)) = \Sigma(W_H(\Omega(\bar{T}))),$$

since the $W$-construction commutes with symmetrization.

The coloured operad $W_H(\Omega(\bar{T}))$ can be described explicitly (Example 6.3.2). The colours of $W_H(\Omega(\bar{T}))$ are the colours of $\Omega(\bar{T})$, i.e., the edges of $T$. Let $\sigma = (e_1, \ldots, e_n; e)$ be an $(n+1)$-tuple of colours of $\Omega(\bar{T})$. If $\Omega(\bar{T})(\sigma) = \emptyset$, then $W_H(\Omega(\bar{T}))(\sigma) = 0$. If $\Omega(\bar{T})(\sigma) \neq \emptyset$, then there is a subtree $T_\sigma$ of $T$ (and a corresponding planar subtree $\bar{T}_\sigma$ of $\bar{T}$) whose leaves are $e_1, \ldots, e_n$ and whose root is $e$. Thus we have that

$$W_H(\Omega(\bar{T}))(e_1, \ldots, e_n; e) = \bigotimes_{f \in \mathrm{inn}(\sigma)} H,$$

where $\mathrm{inn}(\sigma)$ is the set of *inner* edges of $T_\sigma$ (or of $\bar{T}_\sigma$). This last tensor product can be interpreted as the 'space' of assignments of lengths to inner edges in $\bar{T}_\sigma$; it is the unit if $\mathrm{inn}(\sigma)$ is empty.

The composition product in the coloured operad $W_H(\Omega(\bar{T}))$ is given in terms of the $\circ_i$-operations. If $\sigma = (e_1, \ldots, e_n; e)$ and $\rho = (f_1, \ldots, f_m; e_i)$ are two $(n+1)$-tuples of colours, then the composition map

$$\Omega(\bar{T})(e_1, \ldots, e_n; e_0) \otimes \Omega(\bar{T})(f_1, \ldots, f_m; e)$$

$$\downarrow{\circ_i} \tag{7.6}$$

$$\Omega(\bar{T})(e_1, \ldots, e_{i-1}, f_1, \ldots, f_m, e_{i+1}, \ldots, e_n; e)$$

is defined as follows. The trees $\bar{T}_\sigma$ and $\bar{T}_\rho$ are grafted along $e_i$ to form the tree $\bar{T}_\sigma \circ_{e_i} \bar{T}_\rho$, that is again a planar subtree of $\bar{T}$. In fact, $\bar{T}_\sigma \circ_{e_i} \bar{T}_\rho = \bar{T}_{\sigma \circ_i \rho}$, where

$\sigma \circ_i \rho = (e_1, \ldots, e_{i-1}, f_1, \ldots, f_m, e_{i+1}, \ldots, e_n; e_0)$. For the sets of inner edges we have

$$\mathrm{inn}(\sigma \circ_i \rho) = \mathrm{inn}(\sigma) \cup \mathrm{inn}(\rho) \cup \{e_i\}.$$

The composition product in (7.6) is the map

$$
\begin{array}{ccc}
H^{\otimes\mathrm{inn}(\sigma)} \otimes H^{\otimes\mathrm{inn}(\rho)} & \cdots\cdots\cdots\cdots\cdots\rightarrow & H^{\otimes\mathrm{inn}(\sigma\circ_i\rho)} \\
\cong \Big\downarrow & & \Big\downarrow \cong \\
H^{\otimes\mathrm{inn}(\sigma)\cup\mathrm{inn}(\rho)} \otimes I & \xrightarrow{\ \mathrm{id}\otimes 1\ } & H^{\otimes\mathrm{inn}(\sigma)\cup\mathrm{inn}(\rho)} \otimes H,
\end{array}
$$

where $1\colon I \longrightarrow H$ is one of the endpoints of the interval $H$.

This description of the operad $W_H(\Omega(\bar{T}))$ is functorial in the planar tree $T$. In particular, for an inner edge $e$ of $T$, the tree $T/e$ inherits a planar structure $\overline{T/e}$ from $\bar{T}$, and $W_H(\Omega(\overline{T/e})) \longrightarrow W_H(\Omega(\bar{T}))$ is the natural map assigning length 0 to the edge $e$.

*Proof of Theorem 7.4.1.* Let $T$ be a tree in $\Omega$ and $a$ an inner edge of $T$. We need to find an extension to the following diagram:

$$
\begin{array}{ccc}
\Lambda^a[T] & \xrightarrow{\ \varphi\ } & hcN_d(P). \\
\Big\downarrow & \nearrow & \\
\Omega[T] & &
\end{array}
$$

Let $\bar{T}$ be a planar representative of $T$. An extension $\psi\colon \Omega[T] \longrightarrow hcN_d(P)$ corresponds, by adjointness, to a morphism of non-$\Sigma$ operads

$$\hat{\psi}\colon W_H(\Omega(\bar{T})) \longrightarrow P.$$

For each face map $S \longrightarrow T$, the tree $S$ inherits a planar structure $\bar{S}$ from $\bar{T}$, and the given map $\varphi\colon \Lambda^a[T] \longrightarrow hcN_d(P)$ corresponds, again by adjointness, to a map of operads in $\mathcal{E}$,

$$\hat{\varphi}\colon W_H(\Lambda^a[T]) \longrightarrow P,$$

where we view $W_H(\Lambda^a[T])$ as the colimit of operads in $\mathcal{E}$,

$$W_H(\Lambda^a[T]) = \mathrm{colim}\, W(\Omega(\bar{S})) \tag{7.7}$$

over all the faces of $T$ except the one contracting $a$. In other words, $\varphi$ corresponds to a compatible family of maps

$$\hat{\varphi}_S\colon W_H(\Omega(\bar{S})) \longrightarrow P.$$

We will show that there exists an operad map $\hat{\psi}$ extending $\hat{\varphi}_S$ for all faces $S \neq T/a$. Note that the colours of $\Omega(\bar{T})$ are the same as those of the colimit in (7.7), so we have a map $\psi_0 = \varphi_0$ on colours:

$$\psi_0 \colon E(T) \longrightarrow \{\text{Colours of } P\}.$$

If $\sigma = (e_1, \ldots, e_n; e)$ is an $(n+1)$-tuple of edges of $T$ such that $W_H(\Omega(\bar{T})) \neq \emptyset$, and $T_\sigma \subseteq T$ (with $T_\sigma \neq T$), then $T_\sigma$ is contained in an outer face $S$ of $T$. Hence $W_H(\Omega(\bar{T}))(\sigma) = W_H(\Omega(\bar{T}_\sigma))(\sigma) = W_H(\Omega(\bar{S}))(\sigma)$, and we have a map

$$\hat{\varphi}_S(\sigma) \colon W_H(\Omega(\bar{T}))(\sigma) \longrightarrow P(\sigma),$$

given by $\hat{\varphi}_S \colon W_H(\Omega(\bar{S})) \longrightarrow P$. Thus, the only part of the map of operads $\hat{\psi} \colon W_H(\Omega(\bar{T})) \longrightarrow P$ not determined by $\varphi$ is the one when $T_\tau = T$, where $\tau = (e_1, \ldots, e_n; e)$ and $e_1, \ldots, e_n$ are all the input edges of $\bar{T}$ (in the planar order) and $e$ is the output edge. In this case, $\hat{\psi}(\tau)$ has to be a map

$$\hat{\psi} \colon W_H(\Omega(\bar{T}))(\tau) = H^{\otimes i(\tau)} \longrightarrow P(\tau)$$

such that $\hat{\psi}(\sigma) = \hat{\varphi}_S(\sigma)$ if $\sigma \neq \tau$, and together with these $\hat{\psi}(\sigma)$ respects operad composition. The first condition determines $\hat{\psi}(\tau)$ on the subobject of $H^{\otimes i(\tau)}$ which is given by a value 0 on one of the tensor factors marked by an edge $e_i$ *other than* the given $a$. The second condition determines $\hat{\psi}(\tau)$ on the subobject of $H^{\otimes i(\tau)}$ which is given by a value 1 on one of the factors. Thus, if we write 1 for the map $I \rightarrowtail H$ and $\partial H \rightarrowtail H$ for the map $I \coprod I \longrightarrow H$, and define $\partial H^{\otimes k} \rightarrowtail H^{\otimes k}$ by the Leibniz rule (i.e., $\partial(A \otimes B) = \partial(A) \otimes B \cup A \otimes \partial(B)$), then the problem of finding $\hat{\psi}(\tau)$ is the same as finding an extension to the diagram

$$\partial(H^{\otimes \text{inn}(\sigma) \backslash \{a\}} \otimes H) \cup H^{\otimes \text{inn}(\sigma) \backslash \{a\}} \otimes I \longrightarrow P(\tau)$$
$$\downarrow \qquad\qquad \nearrow^{\hat{\psi}(\sigma)}$$
$$H^{\otimes \text{inn}(\sigma) \backslash \{a\}} \otimes H \xrightarrow{\;\cong\;} H^{\otimes \text{inn}(\sigma)}.$$

This extension exists because $P(\tau)$ is fibrant by assumption, and the left-hand map is a trivial cofibration by the pushout-product axiom for monoidal model categories. □

## 7.5 The exponential property

Recall from Theorem 4.2.2 that the category of dendroidal sets is a closed symmetric monoidal category. The main result of this section is that the internal hom of this monoidal structure $\text{Hom}_{dSets}(D, Y)$ is an inner Kan complex if $D$ is normal and $Y$ is inner Kan. It is a consequence of the following result from [MW09], which we quote here without proof:

**Theorem 7.5.1 ([MW09, Propostion 9.2]).** *Let $S$ and $T$ be any two trees in $\Omega$. Then the natural map*

$$\partial\Omega[S] \otimes \Omega[T] \bigcup_{\partial\Omega[S]\otimes\Lambda^e[T]} \Omega[S] \otimes \Lambda^e[T] \longmapsto \Omega[S] \otimes \Omega[T]$$

*is inner anodyne, where $e$ is any inner edge of $T$.*               □

It follows by standard arguments with saturated classes that, if $A \longmapsto B$ is a normal monomorphism and $C \longmapsto D$ is inner anodyne, then

$$A \otimes D \bigcup_{A\otimes C} B \otimes C \longmapsto B \otimes D \qquad (7.8)$$

is again inner anodyne. Using the Hom-$\otimes$ adjunction, one draws the standard conclusions, such as that if $Y \longrightarrow X$ is an inner Kan fibration and $C \longmapsto D$ is inner anodyne, then

$$\operatorname{Hom}(D, Y) \longrightarrow \operatorname{Hom}(C, Y) \times_{\operatorname{Hom}(C,X)} \operatorname{Hom}(D, X) \qquad (7.9)$$

has the right lifting property with respect to normal monomorphisms. If $C \longmapsto D$ is just normal, then (7.9) is an inner Kan fibration. In particular, taking $C = \emptyset$ and $X = 1$ (the terminal object of $dSets$), we obtain the following.

**Theorem 7.5.2 ([MW09, Theorem 9.1]).** *Let $Y$ and $D$ be dendroidal sets and assume that $D$ is a normal dendroidal set. If $Y$ is a (strict) inner Kan complex, then so is $\operatorname{Hom}_{dSets}(D, Y)$.*               □

The result given by Theorem 7.5.1 is also true in *pdSets*. However, for the tensor product to be defined, one has to assume that either $S$ or $T$ is linear (see Remark 4.2.4). The general statement analogous to (7.8) for *pdSets* takes the following form. Let $K \longmapsto L$ be a monomorphism between simplicial sets and let $C \longmapsto D$ be a monomorphism in *pdSets*. Then

$$u_!(K) \otimes D \bigcup_{u_!(K)\otimes C} u_!(L) \otimes C \longmapsto u_!(L) \otimes D$$

is inner anodyne whenever $K \longmapsto L$ or $C \longmapsto D$ is.

# Lecture 8

# Model structures on dendroidal sets

In this lecture, we will discuss the existence of closed Quillen model category structures on the categories of dendroidal sets and of planar dendroidal sets. The existence of these structures was first proved in [CM09a] using results from [Cis06]. Here we present a somewhat different proof based on a result from [Joy02] and a technical result from [CM09a] which involves some combinatorial arguments about trees. In the first section, we explain these two auxiliary results without proof. Assuming them, the proof of the model structure given in the subsequent sections is self contained.

## 8.1 Preliminaries

We begin by recalling Joyal's result. Recall from [Joy02] that, if $X$ is a simplicial inner Kan complex, then $k(X) \subseteq X$ denotes the maximal Kan complex contained in $X$. This simplicial set $k(X)$ has the same vertices as $X$, while an $n$-simplex $x \colon \Delta[n] \longrightarrow X$ belongs to $k(X)$ if and only if, for every $0 \leq i < j \leq n$, the 1-simplex $x_{ij} \colon \Delta[1] \longrightarrow X$ (given by restricting $x$ along the map $\Delta[1] \longrightarrow \Delta[n]$ sending 0 to $i$ and 1 to $j$) represents an isomorphism in the category $\tau(X)$.

Recall also that a map $f \colon \mathcal{D} \longrightarrow \mathcal{C}$ between categories is called a *categorical fibration* if, for any object $d \in \mathcal{D}$ and any isomorphism $\alpha \colon f(d) \longrightarrow c$ in $\mathcal{C}$, there is an isomorphism $\beta \colon d \longrightarrow e$ in $\mathcal{D}$ with $f(\beta) = \alpha$.

**Lemma 8.1.1.** *Let $p \colon Y \longrightarrow X$ be an inner Kan fibration between inner Kan complexes in sSets. Let $\beta \colon y_0 \longrightarrow y_1$ be a morphism in $\tau(Y)$, and let $x \in X_1$ be a 1-simplex in $X$ which represents the morphism $\tau(p)(\beta) \colon x_0 \longrightarrow x_1$ in $\tau(X)$. Then there exists a 1-simplex $y \colon y_0 \longrightarrow y_1$ in $Y$ with $p(y) = x$ which represents $\beta$.*

I. Moerdijk and B. Toën, *Simplicial Methods for Operads and Algebraic Geometry*, Advanced Courses in Mathematics - CRM Barcelona, DOI 10.1007/978-3-0348-0052-5_8, © Springer Basel AG 2010

*Proof.* Given $\beta$ and $x$, choose any 1-simplex $z \in Y_1$ representing $\beta$. Then $p(z)$ represents the same arrow in $\tau(X)$ as $x$. Thus, there is a 'homotopy' $H \in X_2$ which looks like

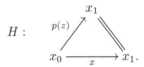

Now use that $p$ is an inner Kan fibration to find a filling of the inner horn

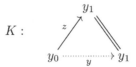

in $Y$ with $p(K) = H$,

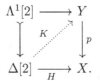

Then $d_1(z) = y$ is the required 1-simplex representing $\beta$ over $x$. □

**Proposition 8.1.2.** *Let $p: Y \longrightarrow X$ be an inner Kan fibration between simplicial inner Kan complexes. If $\tau(Y) \longrightarrow \tau(X)$ is a categorical fibration, $k(Y) \longrightarrow k(X)$ is a Kan fibration.*

*Proof.* This essentially follows from [Joy02, Theorem 2.2]. Indeed, consider a commutative diagram

$$
\begin{array}{ccc}
\Lambda^i[n] & \xrightarrow{\ g\ } & Y \\
\downarrow & & \downarrow{\scriptstyle p} \\
\Delta[n] & \xrightarrow{\ f\ } & X
\end{array}
$$

and suppose that $f$ and $g$ factor through $k(X)$ and $k(Y)$ respectively. If $0 < i < n$, a diagonal lift to the diagram exists by the assumption on $p$. If $i = 0$ and $n > 1$, then a diagonal lift exists by [Joy02, Theorem 2.2], since $g$ is assumed to factor through $k(Y)$. The same is true if $i = n$ by the dual of Joyal's result.

   In each of these cases, the lift $h: \Delta[n] \longrightarrow Y$ factors through $k(Y)$. Indeed, at most one of the 1-dimensional ribbons of $\Delta[n]$ does not belong to $\Lambda^i[n]$, and the image of this ribbon under $h$ must represent an isomorphism in $\tau(Y)$ because all the other ribbons do by assumption. This covers the case $n > 1$. If $n = 1$, then a 'lift up to homotopy' exists by the assumption that $\tau(Y) \longrightarrow \tau(X)$ is a categorical fibration, and this lift can be strictified by Lemma 8.1.1 or its dual version. This concludes the proof of the proposition. □

**Corollary 8.1.3.** *Under the assumptions of Proposition 8.1.2, if $\tau(Y) \longrightarrow \tau(X)$ is an equivalence of categories, then any commutative diagram in sSets*

$$
\begin{array}{ccc}
1+1 & \xrightarrow{\ y\ } & Y \\
\big\downarrow & & \big\downarrow{\scriptstyle p} \\
J & \xrightarrow{\ F\ } & X
\end{array}
$$

*has a diagonal filling.*

*Proof.* Consider the canonical inclusion $\Delta[1] \longrightarrow J$. The map $F$ restricts to a 1-simplex $f\colon \Delta[1] \longrightarrow X$ representing an isomorphism in $\tau(X)$, while $y$ corresponds to two vertices $y_0$ and $y_1$ in $Y$. Since $\tau(Y) \longrightarrow \tau(X)$ is an equivalence of categories, there exists a $g\colon y_0 \longrightarrow y_1$ in $Y_1$ such that $p(y)$ represents the same morphism as $f$ in $\tau(X)$. By Lemma 8.1.1, we can modify $g$ so that $p(g) = f$.

Next, since $F\colon J \longrightarrow X$ and this $g\colon \Delta[1] \longrightarrow Y$ factor through the Kan complexes $k(X)$ and $k(Y)$ respectively, and $\Delta[1] \rightarrowtail J$ is a trivial cofibration in the ordinary model structure on *sSets*, we find a lifting in

$$
\begin{array}{ccc}
\Delta[1] & \longrightarrow & k(Y) \subseteq Y \\
\big\downarrow & \nearrow & \big\downarrow{\scriptstyle p} \\
J & \xrightarrow{\ F\ } & k(X) \subseteq X,
\end{array}
$$

by Proposition 8.1.2. □

We wish to apply Proposition 8.1.2 and Corollary 8.1.3 to simplicial sets arising as homs between (planar) dendroidal sets. To this end, we need the following result from [CM09a], which we state without proof.

**Proposition 8.1.4.** *The following hold:*

(i) *If $X$ is an inner Kan complex in pdSets and $A \rightarrowtail B$ is a monomorphism in pdSets, then the map of simplicial inner Kan complexes*

$$\mathrm{Hom}(B, X) \longrightarrow \mathrm{Hom}(A, X)$$

*induces a categorical fibration*

$$\tau \, \mathrm{Hom}(B, X) \longrightarrow \tau \, \mathrm{Hom}(A, X).$$

(ii) *If $X$ is an inner Kan complex in dSets and $A \rightarrowtail B$ is a monomorphism in dSets where $A$ and $B$ are normal dendroidal sets, then the map of simplicial inner Kan complexes*

$$i^* \operatorname{Hom}_{dSets}(B, X) \longrightarrow i^* \operatorname{Hom}_{dSets}(A, X)$$

*induces a categorical fibration*

$$\tau i^* \operatorname{Hom}_{dSets}(B, X) \longrightarrow \tau i^* \operatorname{Hom}_{dSets}(A, X).$$

*Proof.* Case (ii) is proved in detail in [CM09a], and exactly the same argument shows (i). □

### 8.1.1   Tensor product

Recall from Section 4.1 that the Boardman–Vogt tensor product $P \otimes Q$ of two operads $P$ and $Q$ is defined for *symmetric* operads only. It induces a *symmetric monoidal closed structure* $\otimes$, Hom on *dSets*, completely determined by the definition of the tensor of two representables,

$$\Omega[S] \otimes \Omega[T] = N_d(\Omega(S) \otimes_{BV} \Omega(T)),$$

where $\Omega(S)$ and $\Omega(T)$ are the coloured operads associated to $S$ and $T$.

A trivial but important observation from Lecture 4 is that one does not need the symmetries to describe $P \otimes Q$ if $P$ or $Q$ has unary operations only. Thus, in *pdSets*, one can define

$$u_!(\Delta[n]) \otimes \Omega_p[T],$$

and this makes *pdSets* into a *simplicial category* with tensors and cotensors (see Theorem 4.2.5). The simplicial structure on *pdSets* will alternatively be denoted by

$$S \otimes X \qquad \text{or} \qquad u_!(S) \otimes X.$$

In summary, the Boardman–Vogt tensor product makes *sSets* into a *cartesian closed category*, *pdSets* into a *simplicial category*, and *dSets* into a *symmetric monoidal closed category*. Moreover, the functors $i_!$, $u_!$ and $v_!$ preserve the tensor structure up to isomorphism. In particular, the unit for the monoidal structure in *dSets* is

$$U = i_!(\Delta[0]) = \Omega[\eta],$$

where $\eta$ is the tree with just one edge and no vertices. We will also write

$$U_p = u_!(\Delta[0]) = \Omega_p[\eta].$$

Recall also that $dSets/U = sSets$ and $pdSets/U_p = sSets$.

### 8.1.2  Intervals

We denote by $J$ the simplicial set $N(0 \xleftrightarrow{\sim} 1)$. It has the structure of an interval (see Definition 6.2.1):

$$1 + 1 \rightarrowtail J \longrightarrow 1,$$

so it can be used to describe '$J$-homotopies'. It induces similar intervals $J_p = u_!(J)$ in *pdSets* and $J_d = i_!(J)$ in *dSets*, with structure maps

$$U_p + U_p \rightarrowtail J_p \xrightarrow{\varepsilon} U_p \qquad \text{and} \qquad U + U \rightarrowtail J_d \longrightarrow U,$$

so that we can speak of two maps in *pdSets* being $J_p$-homotopic, and maps in *dSets* being $J_d$-homotopic.

### 8.1.3  Normalization

Recall that a monomorphism $X \rightarrowtail Y$ in *dSets* is called *normal* if, for every tree $T$ in $\Omega$, every non-degenerate element $y \in Y(T)$ which is not in the image of $X(T)$ has a trivial stabilizer $\mathrm{Aut}(T)_y \subseteq \mathrm{Aut}(T)$. An object $X$ is called *normal* if the map $\emptyset \rightarrowtail X$ is a normal monomorphism.

The normal monomorphisms in *dSets* form a saturated class. In fact, it is the saturation of all the boundary inclusions

$$\partial\Omega[T] \rightarrowtail \Omega[T],$$

and also the saturation of the class of images

$$v_! A \rightarrowtail v_! B$$

of monomorphisms $A \rightarrowtail B$ in *pdSets*. By the usual small object argument, every map $X \longrightarrow Y$ of dendroidal sets factors as a normal monomorphism $X \rightarrowtail Z$ and a map $Z \longrightarrow Y$ having the right lifting property with respect to all normal monomorphisms. In particular, when we apply this to $X = \emptyset$, we find for each object $Y$ a normal object $Y_n$ and a cover $Y_n \twoheadrightarrow Y$ having the right lifting property with respect to all normal monomorphisms. We call such a cover $Y_n \twoheadrightarrow Y$ a *normalization* of $Y$.

**Lemma 8.1.5.** *Normalizations are unique up to $J_d$-homotopy equivalence. The standard construction gives for every $Y$ a normalization $Y_n \twoheadrightarrow Y$ with countable fibers.*

*Proof.* Let $p \colon V \longrightarrow Y$ and $q \colon W \longrightarrow Y$ be two normalizations. We find maps $f \colon V \longrightarrow W$ and $g \colon W \longrightarrow V$ over $Y$ by lifting in

Next, lifting in

shows that $f$ and $g$ are mutually inverse up to $J_d$-homotopy.

   To prove the second part, observe that the standard construction produces $Y_n$ as $\bigcup Y^{(i)}$ where $Y^{(0)} = \emptyset$ and $Y^{(i+1)} = Y^{(i)\prime}$; here, for any map $f \colon B \longrightarrow Y$, we write $f' \colon B' \longrightarrow Y$ for the map constructed as the pushout

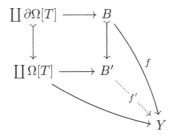

where the sum is over all $T$ and all maps $e$, $e'$ fitting into a commutative square

$$
\begin{array}{ccc}
\partial\Omega[T] & \xrightarrow{\;e'\;} & B \\
\downarrow & & \downarrow{\scriptstyle f} \\
\Omega[T] & \xrightarrow{\;e\;} & Y.
\end{array}
$$

We claim that if $f \colon B \longrightarrow Y$ has countable fibers, then so does $f' \colon B' \longrightarrow Y$. Indeed, fix $R \in \Omega$ and $y \in Y(R)$, and consider the elements $b' \in B'(R)$ with $f'(b') = y$. Such elements $b'$ either (1) lie in $B(R)$, or (2) are given by a triple $(\gamma \colon R \longrightarrow T, e \colon \Omega[T] \longrightarrow Y, e' \colon \partial\Omega[T] \longrightarrow B)$ with $e\gamma = y$, and $e$, $e'$ fitting into a commutative square as above. In this second case, we may assume that $\gamma \colon R \longrightarrow T$ does not factor through the boundary of $\Omega[T]$, because in that case $b' \in B'(R)$ is (represented by) an element of $B(R)$ and we are back in the first case. Now, for the first case $b' \in B(R)$ there are only countably many choices, because $f^{-1}(y)$ is assumed to be countable. For the second case, since we can assume that $\gamma \colon R \longrightarrow T$ does not factor through the boundary, the map $\gamma$ is a composition of degeneracies and an isomorphism, hence it is a (split) epimorphism. So $e$ is uniquely determined by $\gamma$ and the equation $e\gamma = y$. Since there are at most countably many such $\gamma$, there are at most countable many choices $e'$ giving a commutative square as above, by the assumption that $B \longrightarrow Y$ has countable fibers. This proves that $B' \longrightarrow Y$ has countable fibers again.                                    $\square$

Using Lemma 3.4.5, we can see that one can easily pull back or push forward normalizations: given a map $X \longrightarrow Y$ and a normalization $Y_n \longrightarrow\!\!\!\!\!\rightarrow Y$ of $Y$, the pullback $X \times_Y Y_n \longrightarrow X$ is a normalization of $X$; and vice versa, given a normalization $X_n \longrightarrow\!\!\!\!\!\rightarrow X$ of $X$, one obtains one of $Y$ by factoring the composite $X_n \longrightarrow\!\!\!\!\!\rightarrow X \longrightarrow Y$ as a normal monomorphism followed by a map having the right lifting property with respect to all normal monomorphisms.

## 8.2 A Quillen model structure on planar dendroidal sets

In this section we state the main result for planar dendroidal sets, and consider some of its consequences. Let us define a map $f \colon X \longrightarrow Y$ in *pdSets* to be a *weak equivalence* if, for every inner Kan complex $K$, the induced map

$$f^* \colon \tau \operatorname{Hom}(Y, K) \longrightarrow \tau \operatorname{Hom}(X, K) \tag{8.1}$$

is an equivalence of categories. (Recall that *pdSets* is a simplicial category, and that $\operatorname{Hom}(X, K)$ is a quasi-category for any two planar dendroidal sets $X$ and $K$ with $K$ inner Kan.)

**Theorem 8.2.1.** *These weak equivalences are part of a closed model structure on the category of planar dendroidal sets, in which the cofibrations are the monomorphisms.*

*Outline of proof.* The plan of the proof is quite standard, and isolates the difficult parts as Proposition 8.2.5 to be proved later. Recall that a monomorphism between planar dendroidal sets is a map $u \colon A \longrightarrow B$ which induces a one-to-one map of sets $A(T) \longrightarrow B(T)$ for every planar tree $T$ (an object of $\Omega_p$). By the closedness of the model structure, the fibrations are the maps with the right lifting property with respect to the trivial cofibrations. The axioms CM1–CM5 of [Qui69] can now be verified along the following lines:

CM1. The category *pdSets* has all small limits and colimits, and these are calculated 'pointwise', like in any presheaf category.

CM2. (2 out of 3) The 'two out of three' property for weak equivalences follows from the corresponding one for categories.

CM3. (Retracts) The classes of cofibrations and fibrations are clearly closed under retracts. The class of weak equivalences is as well, by functoriality of $f \mapsto f^*$ in (8.1) and the corresponding fact for (weak) equivalences between categories.

CM4. (Lifting) Consider a commutative square

$$\begin{array}{ccc} A & \xrightarrow{\;f\;} & Y \\ {\scriptstyle u}\big\downarrow & & \big\downarrow{\scriptstyle p} \\ B & \xrightarrow{\;g\;} & X, \end{array} \tag{8.2}$$

where $u$ is a cofibration and $p$ is a fibration. If $u$ is also a weak equivalence, then a diagonal filling exists by definition of the fibrations. And if $p$ is also a weak equivalence, we factor $p$ as $p = qv$ where $v$ is a cofibration and $q$ has the right lifting property with respect to all the cofibrations, as in (CM5a) below. Then both $q$ and $v$ are weak equivalences (by CM2 and Proposition 8.2.5). By definition of the fibrations, there is a map $r$ with $rv = \mathrm{id}_Y$ ad $pr = q$. Also, since $q$ has the right lifting property with respect to all the cofibrations, there is a map $h$ with $qh = g$ and $hu = vf$. Then $rh$ is the required lifting in (8.2), as $p(rh) = g$ and $(rh)u = f$.

CM5a. Since, as in any presheaf category, there is a small set of generators for the monomorphisms, the usual small object argument provides for any $f \colon Y \longrightarrow X$ a factorization into a monomorphism $v \colon Y \longrightarrow Z$ and a map $q \colon Z \longrightarrow X$ having the right lifting property with respect to all monomorphisms. Such a map $q$ is evidently a fibration, and we will show separately in Proposition 8.2.5 that it is also a weak equivalence.

CM5b. The main difficulty will be to show that there is a small set of generators for the trivial cofibrations. This will be proved in Section 8.3. The usual small object argument then provides a factorization of an arbitrary map $p \colon Y \longrightarrow X$ into a trivial cofibration $v \colon Y \longrightarrow Z$ followed by a map $q \colon Z \longrightarrow X$ having the right lifting property with respect to the trivial cofibrations, i.e., a fibration. □

**Theorem 8.2.2.** *In the model structure of Theorem 8.2.1, the fibrant objects are exactly the inner Kan complexes.*

*Proof.* In one direction, it is clear that every fibrant object is an inner Kan complex, because each inner horn $\Lambda_p^e[T] \rightarrowtail \Omega_p[T]$ is in fact a weak equivalence, cf. Proposition 8.3.1(i) below. Conversely, let $K$ be an inner Kan complex in *pdSets*, and let $u \colon A \rightarrowtail B$ be a trivial cofibration. By the exponential property (see Section 7.5), both simplicial sets $\mathrm{Hom}(A, K)$ and $\mathrm{Hom}(B, K)$ are inner Kan complexes, and the map $\mathrm{Hom}(B, K) \longrightarrow \mathrm{Hom}(A, K)$ is an inner Kan fibration. Moreover,

$$\tau \,\mathrm{Hom}(B, K) \longrightarrow \tau \,\mathrm{Hom}(A, K)$$

is an equivalence of categories by definition of the weak equivalences, and a fibration between categories by Proposition 8.1.4(i). But any equivalence of categories which is also a fibration is surjective on objects. In this case, this means that $\mathrm{Hom}(B, K) \longrightarrow \mathrm{Hom}(A, K)$ is surjective on vertices, i.e., that any map $A \longrightarrow K$ extends to $B$. This proves that $K$ is fibrant. □

**Corollary 8.2.3.** *There exists a closed model structure on simplicial sets in which the cofibrations are the monomorphisms and the fibrant objects are exactly the quasi-categories.*

*Proof.* Recall that $pdSets/U_p = sSets$, where $U_p = u_!(\Delta[0])$. As for any slice category, the model structure of Theorem 8.2.1 induces one on $pdSets/U_p$, with the 'same' cofibrations, weak equivalences and fibrations. Its fibrant objects are the fibrations $\widehat{X} \longrightarrow U_p$, corresponding under the equivalence $pdSets/U_p = sSets$ to simplicial sets $X$ for which $u_!(X) \longrightarrow u_!(\Delta[0])$ is a fibration in $pdSets$. If $X$ is such a simplicial set, then clearly $X$ is a quasi-category because $u_!$ is fully faithful. And conversely, if $X$ is a quasi-category, then $u_!(X)$ is an inner Kan complex by Proposition 7.1.3, hence fibrant by Theorem 8.2.2. But, in general, if $F$ is a fibrant object and $V \longrightarrow 1$ is any monomorphism into the terminal object, then $F \longrightarrow V$ is a fibration, as one immediately verifies. In particular, $u_!(X) \longrightarrow u_!(\Delta[0])$ is a fibration. □

*Remark* 8.2.4. It follows that the model structure of Corollary 8.2.3 coincides with the one established by Joyal; see [Joy02], [Lur09].

To conclude this section, we prove the following proposition, referred to in the proof of Theorem 8.2.1 above.

**Proposition 8.2.5.** *Any map in pdSets which has the right lifting property with respect to all monomorphisms is a $J_p$-homotopy equivalence (in fact, a $J_p$-deformation retract). Any $J_p$-homotopy equivalence in pdSets is a weak equivalence.*

*Proof.* Let $f \colon Y \longrightarrow X$ be a map having the right lifting property with respect to all monomorphisms. Lifting in

$$
\begin{array}{ccc}
\emptyset & \longrightarrow & Y \\
\downarrow & & \downarrow f \\
X & = & X
\end{array}
$$

shows that $f$ has a section $s \colon X \longrightarrow Y$. Next, using the interval

$$U_p + U_p \rightarrowtail J_p \xrightarrow{\ \varepsilon\ } U_p,$$

the lift in

$$
\begin{array}{ccc}
(U_p + U_p) \otimes Y \bigcup_{(U_p+U_p)\otimes X} J_p \otimes X & \longrightarrow & Y \\
\downarrow & & \downarrow f \\
J_p \otimes X & \longrightarrow & X
\end{array}
$$

shows that $sf$ is $J_p$-homotopic to the identity, by a homotopy relative to $X$.

To prove the second part, suppose that $f \colon Y \rightleftarrows X \colon g$ are mutually $J_p$-homotopy inverse, by homotopies $H_1 \colon J_p \otimes X \longrightarrow X$ between $fg$ and $\mathrm{id}_X$, and $H_2 \colon J_p \otimes Y \longrightarrow Y$ between $gf$ and $\mathrm{id}_Y$. Let $K$ be any inner Kan complex in $pdSets$, and consider the maps of simplicial sets

$$f^* \colon \mathrm{Hom}(X, K) \rightleftarrows \mathrm{Hom}(Y, K) \colon g^*.$$

These are again mutually homotopy inverse for the simplicial interval $J$, by homotopies

$$\widetilde{H}_1 \colon J \times \operatorname{Hom}(X, K) \longrightarrow \operatorname{Hom}(X, K)$$

$$\widetilde{H}_2 \colon J \times \operatorname{Hom}(Y, K) \longrightarrow \operatorname{Hom}(Y, K)$$

defined explicitly in terms of $H_1$ and $H_2$. Applying the functor $\tau \colon sSets \longrightarrow Cat$ to these, we find that $\tau \operatorname{Hom}(X, K) \rightleftarrows \tau \operatorname{Hom}(Y, K)$ are mutually homotopy inverse for the interval $\tau(J) = (0 \overset{\sim}{\longleftrightarrow} 1)$; i.e., they are equivalences of categories, mutually inverse up to natural isomorphism. $\qquad\square$

## 8.3   Trivial cofibrations

In this section, we discuss various properties of the class of trivial cofibrations between planar dendroidal sets. In particular, we show that there is a small set generating them, cf. Proposition 8.3.7.

To begin with, we will construct for each planar dendroidal set $X$ a trivial cofibration

$$X \rightarrowtail X_\infty$$

into an inner Kan complex $X_\infty$. This is done using the familiar small object argument. We let $X_\infty = \varinjlim X_{(n)}$ where $X_{(0)} = X$ and $X_{(n+1)} = X'_{(n)}$, while for any planar dendroidal set $Y$, the planar dendroidal set $Y'$ is constructed as the pushout

$$
\begin{CD}
\coprod \Lambda^e[T] @>>> Y \\
@VVV @VVV \\
\coprod \Omega_p[T] @>>> Y'
\end{CD}
\qquad (8.3)
$$

where the sum is over all planar trees $T$ (objects of $\Omega_p$), all inner edges $e$, and all maps $\Lambda^e[T] \longrightarrow Y$. We summarize the properties of the construction in the following proposition.

**Proposition 8.3.1.** *Let $X$ be any planar dendroidal set and let $X_\infty$ be the planar dendroidal set constructed above.*

(i) *The map $X \longrightarrow X_\infty$ is a trivial cofibration.*

(ii) *For any object $X$ of pdSets, $X_\infty$ is an inner Kan complex.*

(iii) *The construction $X \longmapsto X_\infty$ is functorial. If $X \longrightarrow Y$ is a (trivial) cofibration, so is $X_\infty \longrightarrow Y_\infty$.*

*Proof.* To prove part (i), it is enough to show that $X \longrightarrow X'$ is a trivial cofibration. The map is clearly a monomorphism. To see that it is a weak equivalence, consider an inner Kan complex $K$ and the pullback of simplicial sets

$$\begin{array}{ccc} \operatorname{Hom}(X',K) & \longrightarrow & \operatorname{Hom}(B,K) \\ \downarrow & & \downarrow \\ \operatorname{Hom}(X,K) & \longrightarrow & \operatorname{Hom}(A,K), \end{array}$$

where $A = \coprod \Lambda^e[T]$ and $B = \coprod \Omega_p[T]$ are the coproducts as in (8.3). We claim that $\operatorname{Hom}(B,K) \longrightarrow \operatorname{Hom}(A,K)$ has the right lifting property with respect to all monomorphisms of simplicial sets. Indeed, it is a product of maps of the form $\operatorname{Hom}(\Omega_p[T],K) \longrightarrow \operatorname{Hom}(\Lambda^e[T],K)$, and each of these has it (cf. Section 7.2). Then the same is true for $\operatorname{Hom}(X',K) \longrightarrow \operatorname{Hom}(X,K)$. Hence this map is a homotopy equivalence of simplicial sets, and therefore $\tau \operatorname{Hom}(X',K) \longrightarrow \tau \operatorname{Hom}(X,K)$ is an equivalence of categories.

Part (ii) is clear, because any map $\Lambda^e[T] \longrightarrow X_\infty$ factors through some $X_{(n)}$; hence an extension to $\Omega_p[T]$ exists, mapping to $X_{(n+1)}$.

To prove part (iii), note that it follows from the construction that $X' \longrightarrow Y'$ is a monomorphism whenever $X \longrightarrow Y$ is. Indeed, the pushout for $Y'$ can be viewed as

$$\begin{array}{ccc} A + A^* & \longrightarrow & Y \\ \downarrow & & \downarrow \\ B + B^* & \longrightarrow & Y' \end{array}$$

where $A = \coprod \Lambda^e[T]$ (resp. $B = \coprod \Omega_p[T]$) is the sum over all maps $\Lambda^e[T] \longrightarrow Y$ which factor through $X$, and $A^*$ (resp. $B^*$) is the sum over all such which do not. Then we can construct the pushout in two steps, as

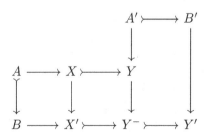

where all squares are pushouts. Since (as in any topos) the pushout of a monomorphism is a monomorphism, the maps $X' \longrightarrow Y^-$ and $Y^- \longrightarrow Y'$ are monomorphisms. It follows by passing to the colimit that $X_\infty \longrightarrow Y_\infty$ is a monomorphism. Moreover, by (i), if $X \longrightarrow Y$ is a weak equivalence, then so is $X_\infty \longrightarrow Y_\infty$. □

**Proposition 8.3.2.** *Any trivial cofibration between inner Kan complexes is a deformation retract.*

*Proof.* Let $u\colon A \longrightarrow B$ be a trivial cofibration between inner Kan complexes. By Theorem 8.2.2, $A$ and $B$ are both fibrant. Thus, lifting in

gives a retraction $r$. To show that $ur \simeq \mathrm{id}_B$ (rel. $A$) we need a lifting in

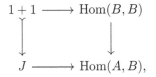

where $\varphi$ is given by $ur$ and $\mathrm{id}_B$ on the summand $(U_p + U_p) \otimes B = B + B$, and by the composition of $\varepsilon \otimes A\colon J_p \otimes A \longrightarrow U_p \otimes A = A$ and $u\colon A \longrightarrow B$ on the summand $J_p \otimes A$. Such a lifting in the previous diagram of planar dendroidal sets is equivalent to a lifting in simplicial sets for

$$
\begin{array}{ccc}
1 + 1 & \longrightarrow & \mathrm{Hom}(B, B) \\
\Big\downarrow & & \Big\downarrow \\
J & \longrightarrow & \mathrm{Hom}(A, B),
\end{array}
$$

and this exists by Corollary 8.1.3.                                        □

**Corollary 8.3.3.** *A map $X \rightarrowtail Y$ in pdSets is a trivial cofibration if and only if $X_\infty \rightarrowtail Y_\infty$ is a deformation retract.*

*Proof.* This follows from Propositions 8.3.1 and 8.3.2, and the 'two out of three' property (CM2).                                        □

**Corollary 8.3.4.** *The class of trivial cofibrations is closed under pushouts.*

*Proof.* Consider a pushout as in the back of the diagram

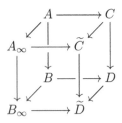

and assume that $A \longrightarrow B$ is a trivial cofibration. Form the square as indicated on the left of the cube, and construct $\widetilde{C}$ and $\widetilde{D}$ so as to make top and bottom pushouts. Then $C \longrightarrow \widetilde{C}$ and $D \longrightarrow \widetilde{D}$ are compositions of pushouts along sums of inner horns $\Lambda^e[T] \longrightarrow \Omega_p[T]$, hence weak equivalences (as in the proof of Proposition 8.3.1(i)). Moreover, since both bottom and top of the cube are pushouts, so is its front. Now $A_\infty \rightarrowtail B_\infty$ is a deformation retract, and these are stable under pushout, so $\widetilde{C} \longrightarrow \widetilde{D}$ is a deformation retract as well, and hence a weak equivalence (cf. Proposition 8.2.5). It follows by 'two out of three' that $C \longrightarrow D$ is a weak equivalence. $\qquad\square$

We will now consider *countable* planar dendroidal sets, i.e., planar dendroidal sets $X$ for which $X(T)$ is countable for each tree $T \in \Omega_p$. Or equivalently, since $\Omega_p$ has (or can be taken to have) only countably many objects, $X$ is countable if and only if its underlying set of dendrices $\coprod X(T)$ is.

**Lemma 8.3.5.** *Let $X$ be any planar dendroidal set.*

(i) *If $X$ is countable, then so is $X_\infty$.*

(ii) *If $U \subseteq X_\infty$ is countable, then there exists a countable $A \subseteq X$ with $U \subseteq A_\infty$.*

*Proof.* Part (i) is clear. For part (ii), we will show by induction on $n$ that, if $V \subseteq X_{(n)}$ is countable, then there exists a countable $B \subseteq X$ with $V \subseteq B_{(n)} \subseteq X_{(n)}$. Assertion (ii) then follows. Indeed, for $U \subseteq X_\infty$ countable, let $U_{(n)} = X_{(n)} \cap U$, and let $A^n \subseteq X$ be countable with the property that $(A^n)_{(n)} \supseteq U_{(n)}$. Let $A = \cup_n A^n$. Then

$$U = \bigcup U_{(n)} \subseteq \bigcup (A^n)_{(n)} \subseteq \bigcup_m \left( \bigcup_n A^n \right)_{(m)} = \bigcup_m A_{(m)} = A_\infty.$$

So, let us prove the inductive statement. For $n = 1$, we have a countable $V \subseteq X'$ and hence $V \cap X$ is countable. Moreover, the dendrices of $V$ which are in $X$ are all contained in a smaller pushout $X \subseteq X_0 \subseteq X'$, constructed as

$$
\begin{array}{ccc}
\coprod_\alpha \Lambda^e[T] & \longrightarrow & X \\
\downarrow & & \downarrow \\
\coprod_\alpha \Omega_p[T] & \longrightarrow & X_0
\end{array}
$$

and involving only the sums over countably many maps $\alpha \colon \Lambda^e[T] \longrightarrow X$ (for varying $T$ and $e$). So, if we let $B$ be the union of $V \cap X$ and the images of these maps, then $B$ is countable and $V \subseteq B'$. Now suppose that the statement has been proved for $X_{(n)}$. Let $V \subseteq X_{(n+1)}$ be countable. By the case $n = 1$ and writing $X_{(n+1)} = X'_{(n)}$, we find a countable $C \subseteq X_{(n)}$ with $C' \supseteq V$. And by the induction hypothesis, we next find a countable $B \subseteq X$ with $B_{(n)} \supseteq C$. Then $B_{(n+1)} \supseteq C' \supseteq V$. This completes the induction and the proof of the lemma. $\quad\square$

**Lemma 8.3.6.** *Let* $u\colon X \rightarrowtail Y$ *be a trivial cofibration, and let* $A \subseteq X$ *and* $B \subseteq Y$ *be countable planar dendroidal subsets. Then there exist larger such subsets* $A \subseteq \overline{A} \subseteq X$ *and* $B \subseteq \overline{B} \subseteq Y$ *such that*

(i) $\overline{A}$ *and* $\overline{B}$ *are both countable,*

(ii) $u$ *restricts to a trivial cofibration* $\overline{A} \longrightarrow \overline{B}$,

(iii) $u^{-1}(\overline{B}) = \overline{A}$.

*Proof.* By Corollary 8.3.3, the extension $u_\infty \colon X_\infty \longrightarrow Y_\infty$ of $u$ is a deformation retract. Let us write $r \colon Y_\infty \longrightarrow X_\infty$ for the retraction and

$$H \colon J \otimes Y_\infty \longrightarrow Y_\infty$$

for the homotopy, with $H(0, -) = \mathrm{id}$ and $H(1, -) = u_\infty r$. Since $A$ and $B$ are countable, there are countable $A' \subseteq X_\infty$ and $B' \subseteq Y_\infty$ for which $r(B) \subseteq A'$ and $u(A) \subseteq B'$ while $H(J \otimes B) \subset B'$. Replacing $A$ and $B$ by $A'$ and $B'$, and repeating this countably many times, we find countable $A_1 \subseteq X_\infty$ and $B_1 \subseteq Y_\infty$ with $A \subseteq A_1$, $B \subseteq B_1$, and such that $u_\infty$, $r$ and $H$ restrict to a deformation retract

$$A_1 \underset{r}{\overset{u_\infty}{\rightleftarrows}} B_1, \qquad H \colon J \otimes B_1 \longrightarrow B_1.$$

Next, by Lemma 8.3.5, there exist countable $U_1 \subseteq X$ and $V_1 \subseteq Y$ for which $A_1 \subseteq (U_1)_\infty$ and $B_1 \subseteq (V_1)_\infty$. We may choose $U_1$ and $V_1$ in such a way that $u^{-1}(V_1) = U_1$, of course. Now repeat this construction, with $U_1$ for $A$ and $V_1$ for $B$, so as to find countable $A_2 \subset X_\infty$ and $B_2 \subseteq Y_\infty$ such that $U_1 \subseteq A_2$ and $V_1 \subseteq B_2$ and the deformation retraction restricts to $A_2 \rightleftarrows B_2$ and $H \colon J \otimes B_2 \longrightarrow B_2$. Next, by Lemma 8.3.5, find $U_2 \subseteq X$ and $V_2 \subseteq Y$ with $A_2 \subseteq (U_2)_\infty$ and $B_2 \subseteq (V_2)_\infty$, and enlarge these if necessary so as to arrange that $u^{-1}(V_2) = U_2$.

Continuing in this way, we build up a ladder

$$
\begin{array}{ccccccccc}
A & \subseteq & A_1 & \subseteq & (U_1)_\infty & \subseteq & A_2 & \subseteq & (U_2)_\infty & \subseteq & \cdots & \subseteq X_\infty \\
& & \downarrow & & \downarrow & & \downarrow & & \downarrow & & & u_\infty \big\updownarrow r \\
B & \subseteq & B_1 & \subseteq & (V_1)_\infty & \subseteq & B_2 & \subseteq & (V_2)_\infty & \subseteq & \cdots & \subseteq Y_\infty
\end{array}
$$

such that $A_k$ is a deformation retract of $B_k$. Let $\overline{A} = \cup U_n$ and $\overline{B} = \cup V_n$. Then $\overline{A}$ and $\overline{B}$ are countable, and $u^{-1}(\overline{B}) = \overline{A}$ because $u^{-1}(V_k) = U_k$. Moreover, $\overline{A}_\infty = (\cup U_n)_\infty = \cup A_n$ and $\overline{B}_\infty = (\cup V_n)_\infty = \cup B_n$. Since each $u \colon A_n \longrightarrow B_n$ is a deformation retract and in particular a weak equivalence, so is $\overline{A}_\infty \longrightarrow \overline{B}_\infty$. By Corollary 8.3.3, $\overline{A} \longrightarrow \overline{B}$ is a weak equivalence.  $\square$

**Proposition 8.3.7.** *The trivial cofibrations between countable objects generate the trivial cofibrations in* pdSets.

*Proof.* Let $u \colon X \rightarrowtail Y$ be a trivial cofibration. Consider the set $E$ of all (non-degenerate) dendrices in $Y$ which are not in the image of $u$, and fix a well ordering on $E$. We are going to construct a decomposition of $u$ into a sequence

$$X = X_0 \rightarrowtail X_1 \rightarrowtail X_2 \rightarrowtail \ldots \rightarrowtail X_\xi \rightarrowtail X_{\xi+1} \rightarrowtail \ldots \rightarrowtail Y, \quad (8.4)$$

where $X_\lambda = \varinjlim_{\xi < \lambda} X_\xi$ at limit ordinals $\lambda$, and each $X_\xi \rightarrowtail X_{\xi+1}$ is a pushout of a trivial cofibration between countable planar dendroidal sets. We construct $X_\xi$ together with $u_\xi \colon X_\xi \longrightarrow Y$ by induction. Let $X_0 = X$ and $u_0 = u \colon X \rightarrowtail Y$. Also, at limit ordinals $\lambda$, let $X_\lambda$ be the colimit as indicated, and let $u_\lambda$ be the map determined by the $u_\xi$ for $\xi < \lambda$ and the universal property of the colimit. To construct $X_{\xi+1}$ from $X_\xi$ and $u_{\xi+1}$ from $u_\xi$, let $y \in Y(T)$ be the first dendrex in the well ordering on $E$ for which $y \notin \mathrm{image}(u_\xi)$. Let $B \subseteq Y$ be a countable planar dendroidal subset containing $y$, and let $A = u_\xi^{-1}(B) \subseteq X_\xi$. Then, by Lemma 8.3.6, there are countable $\overline{A}$ and $\overline{B}$ with $A \subseteq \overline{A} \subseteq X_\xi$ and $B \subseteq \overline{B} \subseteq Y$, such that $u_\xi$ restricts to a trivial cofibration $\overline{A} \longrightarrow \overline{B}$ and $u_\xi^{-1}(\overline{B}) = \overline{A}$. Now form the pushout

$$
\begin{array}{ccc}
\overline{A} & \longrightarrow & X_\xi \\
\Big\downarrow{\scriptstyle\sim} & & \Big\downarrow{\scriptstyle\sim} \\
\overline{B} & \longrightarrow & X_{\xi+1}
\end{array}
$$

and let $u_{\xi+1} \colon X_{\xi+1} \longrightarrow Y$ be the map induced by $u_\xi \colon X_\xi \longrightarrow Y$ and $\overline{B} \subseteq Y$. This map $u_{\xi+1}$ is again a monomorphism, because the square is also a pullback $(u_\xi^{-1}(\overline{B}) = \overline{A})$. This construction stops when no such first $y$ can be found, in which case $X_\xi = Y$. (In particular, the length of the sequence (8.4) is bounded by the length of the well ordering on $E$.) $\qquad\square$

## 8.4 A Quillen model structure on dendroidal sets

In this section, we show how to modify the preceding arguments, so as to obtain a model structure on the category of dendroidal sets. We call a map $f \colon X \longrightarrow Y$ between dendroidal sets a *weak equivalence* if there are normalizations $X_n$ of $X$ and $Y_n$ of $Y$ which fit into a commutative square (cf. Section 8.1.3)

$$
\begin{array}{ccc}
X_n & \xrightarrow{\ f_n\ } & Y_n \\
\Big\downarrow & & \Big\downarrow \\
X & \xrightarrow{\ f\ } & Y,
\end{array}
$$

such that, for any inner Kan complex $K$ in *dSets*, the map

$$\tau i^* \mathrm{Hom}(Y_n, K) \longrightarrow \tau i^* \mathrm{Hom}(X_n, K)$$

is an equivalence of categories. We note that this definition is independent of the choice of the normalization, since they are unique up to homotopy equivalence (see Lemma 8.1.5).

**Theorem 8.4.1.** *These weak equivalences are part of a closed model structure on the category of dendroidal sets, in which the cofibrations are the normal monomorphisms.*

**Proposition 8.4.2.** *In this model structure, the fibrant objects are precisely the inner Kan complexes.*

**Proposition 8.4.3.** *The induced model structure on $dSets/U = sSets$ coincides with the Joyal structure.*

The proof of Theorem 8.4.1 follows exactly the same pattern as that of Theorem 8.2.1. Thus, to establish Theorem 8.4.1, it suffices to prove the following three propositions (analogous to Proposition 8.2.5, Corollary 8.3.4 and Proposition 8.3.7).

**Proposition 8.4.4.** *Let $Y \longrightarrow X$ be a map of dendroidal sets having the right lifting property with respect to all normal monomorphisms. Then $Y \longrightarrow X$ is a weak equivalence.*

**Proposition 8.4.5.** *The class of trivial cofibrations between dendroidal sets is closed under compositions, retracts and pushouts.*

**Proposition 8.4.6.** *The class of trivial cofibrations between dendroidal sets contains a small set of generators.*

*Proof of Proposition* 8.4.4. Consider such a map $f \colon Y \longrightarrow X$. Choose a normalization $r \colon X_n \longrightarrow\!\!\!\!\!\rightarrow X$, and find a lift as in

$$\begin{array}{ccc} \emptyset & \longrightarrow & Y \\ \downarrow & \overset{s}{\nearrow} & \downarrow f \\ X_n & \underset{r}{\longrightarrow\!\!\!\!\!\rightarrow} & X. \end{array}$$

Next, factor the map $s$ as a normal monomorphism $i \colon X_n \rightarrowtail Z$ followed by a map $t \colon Z \longrightarrow Y$ having the right lifting property with respect to all normal monomorphisms, and lift in

$$\begin{array}{ccc} X_n & =\!\!=\!\!= & X_n \\ i \downarrow & \overset{w}{\nearrow} & \downarrow r \\ Z & \underset{ft}{\longrightarrow} & X. \end{array}$$

Then $t \colon Z \longrightarrow Y$ is a normalization of $Y$, and $X_n$ is a retract of $Z$ since $wi = \mathrm{id}_{X_n}$. In fact, it is a deformation retract, i.e., the composition $iw$ is $J_d$-homotopic to the identity on $Z$, as one sees by lifting in

$$
\begin{array}{ccc}
(U + U) \otimes Z \bigcup J_d \otimes X_n & \xrightarrow{\;\varphi\;} & Z \\
\Big\downarrow & {\scriptstyle \varepsilon_Z}\nearrow & \Big\downarrow{\scriptstyle ft} \\
J_d \otimes Z & \xrightarrow[\;\psi\;]{} & X.
\end{array}
$$

Here the map on the left is a normal monomorphism and $ft$ has the right lifting property with respect to all normal monomorphisms. The map $\psi$ is $ft\varepsilon_Z$, and $\varphi$ is given by $\mathrm{id}_Z$ and $iw$ on the two copies of $Z$, and by $i\varepsilon_{X_n}$ on $J_d \otimes X_n$. Thus, $w : Z \rightleftarrows X_n : i$ is a $J_d$-homotopy equivalence, and hence so are the induced maps

$$
\tau i^* \operatorname{Hom}(X_n, K) \rightleftarrows \tau i^* \operatorname{Hom}(Y, K),
$$

i.e., this is an equivalence of categories (as in the proof of Proposition 8.2.5).  $\square$

*Proof of Proposition* 8.4.2. To prove that any fibrant dendroidal set is inner Kan it suffices to show that each inner horn inclusion

$$
\Lambda^e[T] \longrightarrow \Omega[T]
$$

is a trivial cofibration. It is a normal monomorphism between normal objects, and, for any other normal monomorphism $A \longrightarrow B$, the map

$$
\Lambda^e[T] \otimes B \bigcup \Omega[T] \otimes A \longmapsto \Omega[T] \otimes B
$$

is inner anodyne (see Section 7.5). It follows that, for any inner Kan complex $K$, the map $\operatorname{Hom}(\Omega[T], K) \longrightarrow \operatorname{Hom}(\Lambda^e[T], K)$ has the right lifting property with respect to all normal monomorphisms. But then the map

$$
i^* \operatorname{Hom}(\Omega[T], K) \longrightarrow i^* \operatorname{Hom}(\Lambda^e[T], K)
$$

of simplicial sets has the right lifting property with respect to all monomorphisms, hence it is a homotopy equivalence, and therefore $\tau$ maps it to an equivalence of categories.

Conversely, suppose that $X$ is an inner Kan complex. First of all, note that $X \longrightarrow 1$ has the right lifting property with respect to trivial cofibrations between normal objects. Indeed, if $A \rightarrowtail B$ is such a trivial cofibration, then

$$
\operatorname{Hom}(B, X) \longrightarrow \operatorname{Hom}(A, X)
$$

is an inner Kan fibration between inner Kan complexes, and

$$
\tau i^* \operatorname{Hom}(B, X) \longrightarrow \tau i^* \operatorname{Hom}(A, X)
$$

is a categorical fibration and an equivalence of categories, hence surjective on objects. This means that every map $A \longrightarrow X$ extends to $B$:

(This is the same argument as for *pdSets*; cf. Theorem 8.2.2.)

So, it is enough to show that the class of trivial cofibrations is the saturation of the class of trivial cofibrations between normal objects. Take an arbitrary trivial cofibration $u\colon A \rightarrowtail B$. Construct the normalization $q\colon B' \longrightarrow B$ and pull it back to get a normalization $p\colon A' \longrightarrow A$, as in the square

$$
\begin{array}{ccc}
A' & \overset{u'}{\rightarrowtail} & B' \\
{\scriptstyle p}\downarrow & & \downarrow{\scriptstyle q} \\
A & \underset{u}{\rightarrowtail} & B.
\end{array}
$$

Now form the pushout (which will then also be a pullback)

$$
\begin{array}{ccc}
A' & \overset{u'}{\rightarrowtail} & B' \\
{\scriptstyle p}\downarrow & & \downarrow{\scriptstyle r} \\
A & \underset{u}{\rightarrowtail} & P.
\end{array}
\tag{8.5}
$$

By the universal property of this pushout, there is a unique map $s\colon P \longrightarrow B$ with $sr = q$ and $sv = u$. If we can prove that $s$ has the right lifting property with respect to normal monomorphisms, then the proof is finished since we can find a lift

$$
\begin{array}{ccc}
A & \overset{v}{\longrightarrow} & P \\
{\scriptstyle u}\downarrow & \overset{w}{\nearrow} & \downarrow{\scriptstyle s} \\
B & = & B
\end{array}
$$

which makes $u$ a retract of $v$:

$$
\begin{array}{ccccc}
A & = & A & = & A \\
{\scriptstyle u}\downarrow & & \downarrow{\scriptstyle v} & & \downarrow{\scriptstyle u} \\
B & \underset{w}{\longrightarrow} & P & \underset{s}{\longrightarrow} & B.
\end{array}
$$

So $u$ is a retract of a pushout of a trivial cofibration between normal objects (namely $u'$), as desired.

To prove that $s$ has the right lifting property with respect to normal monomorphisms, take a normal monomorphism $U \rightarrowtail V$ and a square

$$
\begin{array}{ccc}
U & \longrightarrow & P \\
\downarrow & & \downarrow s \\
V & \longrightarrow & B.
\end{array}
$$

We may assume that $U$ is normal (in fact, we may assume that $U \rightarrowtail V$ is of the form $\partial\Omega[T] \longrightarrow \Omega[T]$). To find a lift $V \longrightarrow P$, it is enough to lift $U \longrightarrow P$ to $U \longrightarrow B'$:

$$
\begin{array}{ccc}
 & & B' \\
 & \nearrow & \downarrow \\
U & \longrightarrow & P
\end{array}
$$

and then use the right lifting property of $B' \longrightarrow B$:

$$
\begin{array}{ccc}
U & \longrightarrow & B' \\
\downarrow & \nearrow & \downarrow \\
 & & P \\
\downarrow & \nearrow & \downarrow \\
V & \longrightarrow & B.
\end{array}
$$

For this, pull back the pushout square (8.5) along $U \longrightarrow P$ to form the cube

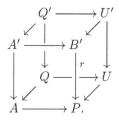

In this diagram, the left, right, bottom and top squares are pullbacks, while the front and back are also pushouts. Then $Q' \longrightarrow Q$ is a normalization and $Q$ is normal (because it lies over $U$). So $Q' \longrightarrow Q$ has a section. But then so does its pushout $U' \longrightarrow U$, say $\sigma : U \longrightarrow U'$. The composition $U \xrightarrow{\sigma} U' \longrightarrow B'$ is the required lift and this finishes the proof. $\qquad\square$

*Proof of Proposition 8.4.3.* The model structure on *dSets* restricts to one on *sSets* (since *sSets* $=$ *dSets*$/U$), in which the cofibrations are all the monomorphisms. Moreover, a map $X \longrightarrow Y$ of simplicial sets is a weak equivalence in this model structure if and only if, for any dendroidal inner Kan complex, the map

$$
\tau i^* \operatorname{Hom}(i_! Y, K) \longrightarrow \tau i^* \operatorname{Hom}(i_! X, K)
$$

is an equivalence of categories. But $i^* \operatorname{Hom}(i_! Y, K) = i^* \operatorname{Hom}(Y, i^* K)$, and $i^*(K)$ is an inner Kan complex of simplicial sets; conversely, every inner Kan complex $F$ of simplicial sets is of this form, for $K = i_!(F)$. So the weak equivalences are also the same as in the Joyal structure.                                                    $\square$

**Lemma 8.4.7.** *Any trivial cofibration between normal inner Kan complexes is a deformation retract.*

*Proof.* This is proved in much the same way as Proposition 8.3.2, except that we cannot use that inner Kan complexes are fibrant. Let $u \colon A \rightarrowtail B$ be a trivial cofibration, and assume that $A$ and $B$ are both inner Kan as well as normal. Consider the map

$$u^* \colon \operatorname{Hom}(B, A) \longrightarrow \operatorname{Hom}(A, A).$$

This is a map between dendroidal inner Kan complexes, as well as an inner Kan fibration. Then $i^*(u^*) \colon i^* \operatorname{Hom}(B, A) \longrightarrow i^* \operatorname{Hom}(A, A)$ is an inner Kan fibration of simplicial sets between quasi-categories. Thus $\tau$ maps it to a fibration of categories (see Proposition 8.1.4(ii)) as well as to an equivalence of categories. But then

$$\tau i^* \operatorname{Hom}(B, A) \longrightarrow \tau i^* \operatorname{Hom}(A, A)$$

must be surjective on objects, which means that there is a retraction $r \colon B \longrightarrow A$ with $ru = \operatorname{id}_A$. Next, to see that $ur$ is homotopic (rel. $A$) to the identity on $B$, we need to lift in

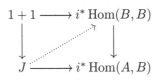

By adjunction, finding such a lift corresponds to finding a lift in the square

$$
\begin{array}{ccc}
1 + 1 & \longrightarrow & i^* \operatorname{Hom}(B, B) \\
\downarrow & \nearrow & \downarrow \\
J & \longrightarrow & i^* \operatorname{Hom}(A, B)
\end{array}
$$

in the category of simplicial sets. Such a lifting exists by Corollary 8.1.3.        $\square$

For a dendroidal set, we again write $X \longrightarrow X_\infty$ for the object obtained by iterated pushouts along sums of inner horn inclusions as in Section 8.3. So $X_\infty$ is an inner Kan complex, and $X \longrightarrow X_\infty$ is inner anodyne. Moreover, the construction is functorial. Also note that $X \rightarrowtail X_\infty$ is normal, so $X_\infty$ is normal whenever $X$ is.

**Lemma 8.4.8.** *For any normal object $X$, the map $X \longrightarrow X_\infty$ is a trivial cofibration.*

*Proof.* This claim is proved exactly as Proposition 8.3.1(i). In fact, the argument shows that every inner anodyne extension between normal objects is a trivial cofibration. □

**Corollary 8.4.9.** *A cofibration (i.e., a normal monomorphism) $u\colon X \rightarrowtail Y$ is trivial if and only if it fits into a diagram*

$$
\begin{array}{ccccc}
X & \xleftarrow{\;p\;} & X_n & \xrightarrow{\;a\;} & (X_n)_\infty \\
{\scriptstyle u}\downarrow & & {\scriptstyle v}\downarrow & & \downarrow{\scriptstyle w} \\
Y & \xleftarrow{\;q\;} & Y_n & \xrightarrow{\;b\;} & (Y_n)_\infty
\end{array}
$$

*where $p$ and $q$ are normalizations of $X$ and $Y$, $a$ and $b$ are inner anodyne, $u$ and $v$ are normal monomorphisms, and $w$ is a deformation retract.*

*Proof.* If $u$ is a trivial cofibration, one can construct a diagram as indicated. Then $w$ is a deformation retract by Lemma 8.4.7.

Conversely, suppose that we have a diagram as above. Then $p$ and $q$ are weak equivalences by Proposition 8.4.4, and $a$ and $b$ are so as well by the remark just made. Since any deformation retract is a weak equivalence, it follows that $u$ is a weak equivalence by the 'two out of three' property. □

*Proof of Proposition 8.4.5.* By functoriality of normalization, it is clear that the trivial cofibrations are closed under retracts and composition. For the case of pushouts, consider a pushout diagram

$$
\begin{array}{ccc}
A & \longrightarrow & C \\
\downarrow & & \downarrow \\
B & \longrightarrow & D
\end{array}
\tag{8.6}
$$

where $A \rightarrowtail B$ is a trivial cofibration. Take a normalization $D_n \twoheadrightarrow D$ and construct a cube

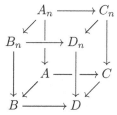

of which all vertical faces are pullback squares. Since, in a topos, pushouts are preserved by pullbacks, the top face is again a pushout. Since $A_n \twoheadrightarrow A$, $B_n \twoheadrightarrow B$ and $C_n \twoheadrightarrow C$ are all normalizations, this reduces the problem to the case where all objects in the square (8.6) are normal. So we assume this from now on. Construct

$A \rightarrowtail A_\infty$ and $B \rightarrowtail B_\infty$, and form another cube in which top and bottom are pushouts, so that the front is automatically a pushout as well:

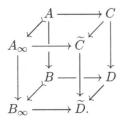

Then $C \longrightarrow \widetilde{C}$ and $D \longrightarrow \widetilde{D}$ are inner anodyne maps between normal objects, hence weak equivalences. Moreover, $A_\infty \longrightarrow B_\infty$ is a deformation retract by Lemma 8.4.7, and hence so is its pushout $\widetilde{C} \longrightarrow \widetilde{D}$. Thus $\widetilde{C} \longrightarrow \widetilde{D}$ is a weak equivalence. It follows that $C \longrightarrow D$ is also a weak equivalence. □

Finally, we prove Proposition 8.4.6. By the argument of Proposition 8.3.7, it suffices to show the analog of Lemma 8.3.6:

**Lemma 8.4.10.** *Let* $u \colon X \rightarrowtail Y$ *be a trivial cofibration between dendroidal sets, and let* $A \subseteq X$, $B \subseteq Y$ *be countable subsets. Then there are larger countable subsets* $A \subseteq \overline{A} \subseteq X$ *and* $B \subseteq \overline{B} \subseteq Y$ *such that*

(i) $\overline{A}$ *and* $\overline{B}$ *are both countable,*

(ii) $u$ *restricts to a trivial cofibration* $\overline{A} \longrightarrow \overline{B}$,

(iii) $u^{-1}(\overline{B}) = \overline{A}$.

*Proof.* Take $A \subseteq X \overset{u}{\rightarrowtail} Y \supseteq B$ as stated in the lemma. Let $p \colon X_n \twoheadrightarrow X$ and $q \colon Y_n \twoheadrightarrow Y$ be the functorial normalizations with countable fibers (cf. Section 8.1.3). Let $A_0 = A$ and $B_0 = B \cup u(A_0)$, so that $u$ maps $A_0$ into $B_0$. We have normal covers $p^{-1}(A_0) \longrightarrow A_0$ and $q^{-1}(B_0) \longrightarrow B_0$, and $p^{-1}(A_0) \subseteq X_n$ and $q^{-1}(B_0) \subseteq Y_n$ are both countable. Using Corollary 8.4.9 and the argument in the proof of Lemma 8.3.6, we find that there exist countable $\widetilde{A}_1$ and $\widetilde{B}_1$ with

$$ p^{-1}(A_0) \subseteq \widetilde{A}_1 \subseteq X_n, \qquad q^{-1}(B_0) \subseteq \widetilde{B}_1 \subseteq Y_n, $$

and such that $X_n \longrightarrow Y_n$ restricts to a trivial cofibration $\widetilde{A}_1 \longrightarrow \widetilde{B}_1$. Let $B_1 = q(\widetilde{B}_1) \supseteq B_0$ be the image of $\widetilde{B}_1$, and let $A_1 = p(\widetilde{A}_1) \cup u^{-1}(B_1)$. Then $A_1$ and $B_1$ are countable again. So, by the same argument, $p^{-1}(A_1)$ and $q^{-1}(B_1)$ are contained in countable $\widetilde{A}_2 \subseteq X_n$ and $\widetilde{B}_2 \subseteq Y_n$ such that $X_n \longrightarrow Y_n$ restricts to a weak equivalence $\widetilde{A}_2 \longrightarrow \widetilde{B}_2$. Now let $B_2 = q(\widetilde{B}_2)$ and $A_2 = p(\widetilde{A}_2) \cup u^{-1}(B_2)$, and so on. In this way, we build a ladder of countable dendroidal sets

$$
\begin{array}{ccccccccc}
A_0 & \rightarrowtail & A_1 & \rightarrowtail & A_2 & \rightarrowtail & \cdots & \rightarrowtail & X \\
\downarrow{\scriptstyle u} & & \downarrow & & \downarrow & & & & \downarrow \\
B_0 & \rightarrowtail & B_1 & \rightarrowtail & B_2 & \rightarrowtail & \cdots & \rightarrowtail & Y
\end{array}
$$

for which the normal covers $p^{-1}(A_i)$ and $q^{-1}(B_i)$ are interpolated by trivial cofibrations $\tilde{A}_i \longrightarrow \tilde{B}_i$,

$$
\begin{array}{ccccccccc}
p^{-1}(A_0) & \rightarrowtail & \tilde{A}_1 & \rightarrowtail & p^{-1}(A_1) & \rightarrowtail & \tilde{A}_2 & \rightarrowtail & \cdots & \rightarrowtail & X_n \\
\downarrow & & \downarrow{\scriptstyle \sim} & & \downarrow & & \downarrow{\scriptstyle \sim} & & & & \downarrow \\
q^{-1}(B_0) & \rightarrowtail & \tilde{B}_1 & \rightarrowtail & q^{-1}(B_1) & \rightarrowtail & \tilde{B}_2 & \rightarrowtail & \cdots & \rightarrowtail & Y_n.
\end{array}
$$

It follows that if we let $\overline{A} = \cup A_n \subseteq X$ and $\overline{B} = \cup B_n \subseteq Y$, then $p^{-1}(\overline{A}) \longrightarrow q^{-1}(\overline{B})$ is a trivial cofibration, and hence so is $\overline{A} \longrightarrow \overline{B}$. Moreover, $u^{-1}(\overline{B}) = \overline{A}$ by construction. □

*Proof of Theorem* 8.4.1. Lemma 8.4.10 together with the argument of Proposition 8.3.7 proves Proposition 8.4.6 and completes the proof. □

# Bibliography

[BM06]  C. Berger and I. Moerdijk, The Boardman–Vogt resolution of operads in monoidal model categories, *Topology* **45** (2006), no. 5, 807–849.

[BM07]  C. Berger and I. Moerdijk, Resolution of coloured operads and rectification of homotopy algebras, in: *Categories in Algebra, Geometry and Mathematical Physics (Street Festschrift)*, Contemp. Math., vol. 431, Amer. Math. Soc., Providence, 2007, 31–58.

[BM08]  C. Berger and I. Moerdijk, On an extension of the notion of Reedy category, preprint, arXiv:0809.3341v1, 2008.

[BV73]  J. M. Boardman and R. M. Vogt, *Homotopy Invariant Algebraic Structures on Topological Spaces*, Lecture Notes in Math., vol. 347, Springer-Verlag, Berlin, Heidelberg, 1973.

[BK72]  A. K. Bousfield and D. M. Kan, *Homotopy Limits, Completions and Localizations*, Lecture Notes in Math., vol. 304, Springer-Verlag, Berlin, Heidelberg, 1972.

[Cis06]  D.-C. Cisinski, Les préfaisceaux comme modèles des types d'homotopie, *Astérisque* **308** (2006).

[CM09a]  D.-C. Cisinski and I. Moerdijk, Dendroidal sets as models for homotopy operads, preprint, arXiv:0902.1954v1, 2009.

[CM09b]  D.-C. Cisinski and I. Moerdijk, Dendroidal Segal spaces and $\infty$-operads, in preparation.

[CP86]  J. M. Cordier and T. Porter, Vogt's theorem on categories of homotopy coherent diagrams, *Math. Proc. Cambridge Philos. Soc.* **100** (1986), 65–90.

[Hir03]  P. S. Hirschhorn, *Model Categories and Their Localizations*, Math. Surveys and Monographs, vol. 99, Amer. Math. Soc., Providence, 2003.

[Hov99]  M. Hovey, *Model Categories*, Math. Surveys and Monographs, vol. 63, Amer. Math. Soc., Providence, 1999.

[Joy02]  A. Joyal, Quasi-categories and Kan complexes, *J. Pure Appl. Algebra* **175** (2002), no. 1–3, 207–222.

[Kel82] G. M. Kelly, *Basic Concepts of Enriched Category Theory*, London Math. Soc. Lecture Notes, vol. 64, Cambridge University Press, Cambridge, 1982.

[Lei04] T. Leinster, *Higher Operads, Higher Categories*, London Math. Soc. Lecture Notes, vol. 298, Cambridge University Press, Cambridge, 2004.

[Lur09] J. Lurie, *Higher Topos Theory*, Annals of Math. Studies, vol. 170, Princeton University Press, Princeton, 2009.

[Mac71] S. Mac Lane, *Categories for the Working Mathematician*, Graduate Texts in Math., vol. 5, Springer-Verlag, New York, 1971.

[May72] J. P. May, *The Geometry of Iterated Loop Spaces*, Lecture Notes in Math., vol. 271, Springer-Verlag, Berlin, Heidelberg, 1972.

[MW07] I. Moerdijk and I. Weiss, Dendroidal sets, *Algebr. Geom. Topol.* **7** (2007), 1441–1470.

[MW09] I. Moerdijk and I. Weiss, On inner Kan complexes in the category of dendroidal sets, to appear in *Adv. Math.*

[Qui67] D. G. Quillen, *Homotopical Algebra*, Lecture Notes in Math., vol. 43, Springer-Verlag, Berlin, Heidelberg, 1967.

[Qui69] D. G. Quillen, Rational homotopy theory, *Ann. of Math.* **90** (1969), no. 2, 205–295.

[Ree74] C. L. Reedy, Homotopy theory of model categories, unpublished manuscript, 1974, available at http://www.math.mit.edu/~psh/reedy.pdf.

# Part II

# Simplicial Presheaves and Derived Algebraic Geometry

Bertrand Toën

# Lecture 1

# Motivation and objectives

The purpose of this first lecture is to present some motivations for derived algebraic geometry and to describe the objectives of this series of lectures. I will start with a brief review of the notion of moduli problems and moduli spaces. In a second part, I will present the particular example of the moduli problem of linear representations of a discrete group. The study of this example will show the importance of two useful constructions to produce and understand moduli spaces: intersections (or, more generally, fiber products) and group quotients (or, more generally, quotients by groupoids). As many algebraic constructions, these fail to be *exact* in some sense, and possess derived versions. This will provide motivation for derived algebraic geometry, which is a geometrico-algebraic setting in which these derived versions exist and are well behaved.

We warn the reader that this section is highly informal and that several notions and ideas will be explained more formally during the lectures.

## 1.1 The notion of moduli spaces

The main object studied in algebraic geometry are schemes (or, more generally, algebraic spaces —these notions will be redefined later). They often appear as solutions to *moduli functors* (or, equivalently, *moduli problems*), which intuitively means that their points classify certain geometrico-algebraic objects (e.g., a scheme whose points are in one-to-one correspondence with algebraic subvarieties of the projective space $\mathbb{P}^n$). More precisely, we are often given a *moduli functor*

$$F \colon Comm \longrightarrow Set$$

from the category of commutative rings to the category of sets. Each set $F(A)$ has to be thought of as a set of families of objects parametrized by the scheme Spec $A$ —we will later see many examples. When it exists, a scheme $X$ is then called a *moduli space* for $F$ (or a *solution to the moduli problem $F$*; we also say that the

I. Moerdijk and B. Toën, *Simplicial Methods for Operads and Algebraic Geometry*, Advanced Courses in Mathematics - CRM Barcelona, DOI 10.1007/978-3-0348-0052-5_9, © Springer Basel AG 2010

scheme $X$ *represents* $F$ or that $F$ is *representable* by $X$) if there are functorial bijections

$$F(A) \simeq \mathrm{Hom}(\mathrm{Spec}\,A, X).$$

An important special case is when $X$ is an affine scheme, say $\mathrm{Spec}\,B$. Then $X$ represents $F$ if and only if there are functorial isomorphisms

$$F(A) \simeq \mathrm{Hom}(B, A),$$

where $\mathrm{Hom}(B, A)$ is the set of ring morphisms from $B$ to $A$.

We mention here a very basic, but fundamental example of a moduli space, namely the projective space $\mathbb{P}^n$ —of course, more elaborate examples will be given in these notes. We define a functor $P_n \colon Comm \longrightarrow Set$ as follows. For $A \in Comm$, we let $P_n(A)$ be the set of sub-$A$-modules $M \subset A^{n+1}$ such that the quotient $A^{n+1}/M$ is projective of rank 1 over $A$ (i.e., an invertible $A$-module). For a morphism $A \longrightarrow B$ in $Comm$, the application $F(A) \longrightarrow F(B)$ sends $M \subset A^{n+1}$ to $M \otimes_A B \subset B^{n+1}$. Note here that, as $A^{n+1}/M$ is projective, $M$ is a direct factor in $A^{n+1}$, and thus $M \otimes_A B$ is a sub-$A$-module of $B^{n+1}$. This defines the functor $P_n$. It is well known that this functor is representable by a scheme, which is denoted by $\mathbb{P}^n$ and called the projective space of dimension $n$.

The notion of moduli space is extremely important for at least two reasons:

1.  A good geometric understanding of the moduli space of a given moduli problem can be considered as a step towards a solution to the corresponding classification problem. For instance, a good enough understanding of the moduli space of algebraic curves could be understood as a solution to the problem of classifying algebraic curves.

2.  The notion of moduli problems is a rich source to construct new and interesting schemes. Indeed, the fact that a given scheme $X$ is the solution to a moduli problem often makes its geometry rather rich. Typically, the scheme will have interesting subschemes corresponding to objects satisfying certain additional properties.

## 1.2   Construction of moduli spaces: one example

For a given moduli problem $F \colon Comm \longrightarrow Set$, the question of the existence of a moduli space is never an easy question. There are two general strategies to prove the existence of such a moduli space, either by applying the so-called *Artin's representability theorem*, or by a more direct approach consisting of constructing the moduli space explicitly. The first approach is the most powerful to prove the existence, but the second one is often needed to have a better understanding of the moduli space itself (e.g., to prove that it satisfies some further properties). In this section we will study the particular example of the moduli problem of linear representations of a discrete group, and will try to construct the corresponding moduli space by a direct approach. This example is chosen so that the moduli space does not exist, which is most often the case, but still the approach to the construction that we present is right, at least when it is done in the context of derived algebraic geometry —we will of course come back to this fundamental example later on, when the techniques of derived stacks are at our disposal.

So let $\Gamma$ be a group, that will be assumed to be finitely presented. We want to study finite-dimensional linear representations of $\Gamma$, and for this we are looking for a moduli space of those. We start by defining a moduli functor

$$R(\Gamma) \colon Comm \longrightarrow Set,$$

sending a commutative ring $A$ to the set of isomorphism classes of $A[\Gamma]$-modules whose underlying $A$-module is projective and of finite type over $A$. As projective $A$-modules of finite type correspond to vector bundles on the scheme $\operatorname{Spec} A$, $R(\Gamma)(A)$ can also be identified with the set of isomorphism classes of vector bundles on $\operatorname{Spec} A$ endowed with an action of $\Gamma$. For a morphism of commutative rings $A \longrightarrow A'$, we have a base-change functor $- \otimes_A A'$ from $A[\Gamma]$-modules to $A'[\Gamma]$-modules, which induces a function

$$R(\Gamma)(A) \longrightarrow R(\Gamma)(A').$$

This defines the moduli functor $R(\Gamma)$.

The strategy to try to construct a solution to this moduli problem is to start by studying a *framed* (or *rigidified*) version of it. We introduce, for any integer $n$, an auxiliary moduli problem $R'_n(\Gamma)$, whose value at a commutative ring $A$ is the set of group morphisms $\Gamma \longrightarrow Gl_n(A)$. We set $R'(\Gamma) = \coprod_n R'_n(\Gamma)$ and define a morphism (i.e., a natural transformation of functors)

$$\pi \colon R'(\Gamma) \longrightarrow R(\Gamma),$$

sending a morphism $\rho \colon \Gamma \longrightarrow Gl_n(A)$ to the $A$-module $A^n$ together with the action of $\Gamma$ defined by $\rho$. At this point we would like to argue in two steps:

1. The moduli functor $R'(\Gamma)$ is representable.

2. The moduli functor $R(\Gamma)$ is the disjoint union (for all $n$) of the quotients of the schemes $R'_n(\Gamma)$ by the group schemes $Gl_n$.

For step 1, we write a presentation of $\Gamma$ by generators and relations

$$\Gamma \simeq \langle g_1, \ldots, g_m \rangle / \langle r_1, \ldots, r_p \rangle.$$

From this presentation, we deduce the existence of a cartesian square of moduli functors

$$
\begin{array}{ccc}
R'_n(\Gamma) & \longrightarrow & Gl_n^m \\
\downarrow & & \downarrow \\
\{1\} & \longrightarrow & Gl_n^p,
\end{array}
$$

where $Gl_n$ is the functor $A \longmapsto Gl_n(A)$. The functor $Gl_n$ is representable by an affine scheme. Indeed, if we set

$$C_n = \mathbb{Z}[T_{i,j}][\text{Det}(T_{i,j})^{-1}],$$

where $T_{i,j}$ are formal variables with $1 \leq i, j \leq n$, then the affine scheme $\text{Spec } C_n$ represents the functor $Gl_n$. This implies that $Gl_n^r$ is also representable by an affine scheme for any integer $r$, precisely $\text{Spec}\,(C_n^{\otimes r})$. And, finally, we see that $R'_n(\Gamma)$ is representable by the affine scheme

$$\text{Spec}\,(C_n^{\otimes m} \otimes_{C_r^{\otimes p}} \mathbb{Z}).$$

This sounds good, but an important observation here is that in general $C_n^{\otimes m}$ is not a flat $C_r^{\otimes p}$-algebra, and thus that the tensor product $C_n^{\otimes m} \otimes_{C_r^{\otimes p}} \mathbb{Z}$ is not well behaved from the point of view of homological algebra. Geometrically, this is related to the fact that $R'_n(\Gamma)$ is the intersection of two subschemes in $Gl_n^m \times Gl_n^p$, namely the graph of the morphism $Gl_n^m \longrightarrow Gl_n^p$ and $Gl_n^m \times \{1\}$, and that these two subschemes are not in general position. A direct consequence of this is the fact that the scheme $R'_n(\Gamma)$ can be badly singular at certain points, precisely the points for which the above intersection is not transversal. Another bad consequence is that the tangent complex (a derived version of the tangent space —we will review this notion later in the course) is not easy to compute for the scheme $R'_n(\Gamma)$. The main philosophy of derived algebraic geometry is that the tensor product $C_n^{\otimes m} \otimes_{C_r^{\otimes p}} \mathbb{Z}$ should be replaced by its derived version $C_n^{\otimes m} \otimes^L_{C_r^{\otimes p}} \mathbb{Z}$, which also encodes the higher Tor's

$$\text{Tor}_*^{C_r^{\otimes p}}(C_n^{\otimes m}, \mathbb{Z}),$$

for instance by considering simplicial commutative rings. Of course, $C_n^{\otimes m} \otimes^L_{C_r^{\otimes p}} \mathbb{Z}$ is no longer a commutative ring, and thus the notion of scheme should be extended in order to be able to consider objects of the form "$\text{Spec } A$", where $A$ is now a simplicial commutative ring.

**Exercise 1.2.1.** Suppose that $\Gamma = \mathbb{Z}^2$, presented by the standard presentation $\Gamma = \langle g_1, g_2 \rangle / \langle [g_1, g_2] \rangle$. Show that the morphism $C_n \longrightarrow C_n^{\otimes 2}$ is indeed a non-flat morphism.

We now consider step 2. The functor $Gl_n \colon Comm \longrightarrow Set$ sending $A$ to $Gl_n(A)$ is a group object (in the category of functors) and acts naturally on $R'_n(\Gamma)$. For a given $A \in Comm$, the action of $Gl_n(A)$ on $R'_n(\Gamma)(A) = \text{Hom}(\Gamma, Gl_n(A))$ is the one induced by the conjugation action of $Gl_n(A)$ on itself. The morphism $R'_n(\Gamma) \longrightarrow R(\Gamma)$ is equivariant for this action and thus factorizes as a morphism

$$R'_n(\Gamma)/Gl_n \longrightarrow R(\Gamma).$$

We thus obtain a morphism of functors

$$\coprod_n R'_n(\Gamma)/Gl_n \longrightarrow R(\Gamma).$$

Intuitively, this morphism should be an isomorphism, and in fact it is close to being one. It is a monomorphism, but it is not an epimorphism because not every projective $A$-module of finite type is free. However, up to a localization for the Zariski topology on $\operatorname{Spec} A$, this is the case, and therefore we see that the above morphism is an epimorphism in the sense of sheaf theory.

In other words, this morphism is an isomorphism if the left-hand side is understood as the *quotient sheaf* with respect to the Zariski topology on the category $Comm^{\mathrm{op}}$. This sounds like a good situation, as both functors $R'_n(\Gamma)$ and $Gl_n$ are representable by affine schemes. However, the quotient sheaf of an affine scheme by the action of an affine group scheme is in general not a scheme when the action has fixed points. It is, for instance, not so hard to see that the quotient sheaf $\mathbb{A}^1/(\mathbb{Z}/2)$ is not representable by any scheme (here $\mathbb{A}^1 = \operatorname{Spec} \mathbb{Z}[T]$ is the affine line and the action is induced by the involution $T \longmapsto -T$; see Ex. 1.2.2). In our situation, the action of $Gl_n$ on $R'_n(\Gamma)$ has many fixed points, since, for a given $A \in Comm$, the stabilizer of a given morphism $\Gamma \longrightarrow Gl_n(A)$ is precisely the group of automorphisms of the corresponding $A[\Gamma]$-module.

We see here that the reason for the non-representability of the quotients $R'_n(\Gamma)/Gl_n$ is the existence of non-trivial automorphism groups. Here the philosophy is the same as for the previous point: the quotient construction is not exact in some sense, and should be derived. The derived quotient of a group $G$ acting on a set $X$ is the groupoid $[X/G]$, whose objects are the points of $X$ and whose morphisms from $x$ to $y$ are the elements $g \in G$ such that $gx = y$. The set of isomorphism classes of objects in $[X/G]$ is the usual quotient $X/G$, but the derived quotient $[X/G]$ also remembers the stabilizers of the action in the automorphism groups of $[X/G]$. This suggests that the right thing to do is to replace $\coprod_n R'_n(\Gamma)/Gl_n$ by the more involved construction $\coprod_n [R'_n(\Gamma)/Gl_n]$, which is now a functor from $Comm$ to the category of groupoids rather than the category of sets.

In the same way, this suggests that the functor $R(\Gamma)$ should rather be replaced by $\underline{R}(\Gamma)$, sending a commutative ring $A$ to the whole groupoid of $A[\Gamma]$-modules whose underlying $A$-module is projective and of finite type. We see here that we need again to extend the notion of scheme in order to be able to find a geometric object representing $\underline{R}(\Gamma)$, as the functor represented by a scheme is always set-valued by definition.

**Exercise 1.2.2.** Let $\mathbb{Z}/2$ act on the scheme $\mathbb{A}^1 = \operatorname{Spec} \mathbb{Z}[T]$ by $T \longmapsto -T$. We denote by $F \colon Comm \longrightarrow Set$ the functor represented by $\mathbb{A}^1$.

1. Show that the quotient of $\mathbb{A}^1$ by $\mathbb{Z}/2$ exists in the category of affine schemes and is isomorphic to $\operatorname{Spec}(\mathbb{Z}[T]^{\mathbb{Z}/2}) \simeq \mathbb{A}^1$.

2. Suppose now that we are given a Grothendieck topology $\tau$ on $Comm^{\mathrm{op}} = Aff$, the category of affine schemes. Let $F_0 = F/(\mathbb{Z}/2)$ be the quotient sheaf for this topology. Prove that the natural morphism $F \times \mathbb{Z}/2 \longrightarrow F \times_{F_0} F$ is an epimorphism of sheaves.

3. Show that, if the topology $\tau$ is sub-canonical, then $F_0$ is represented by an affine scheme if and only if it is represented by $\mathbb{A}^1$.

4. Assume now that $\tau$ is the ffp (flat and finitely presented) topology. Use (2) and (3) to show that $F_0$ cannot be represented by an affine scheme.

## 1.3   Conclusions

We arrive at the conclusions of this first lecture. The fundamental objects of algebraic geometry are functors

$$Comm \longrightarrow Set.$$

However, we have seen that certain constructions on rings (tensor products), or on sets (quotients), are not exact and should rather be derived in order to be better behaved. Deriving the tensor product for commutative rings forces us to introduce simplicial commutative rings, and deriving quotients forces us to introduce groupoids (when it is a quotient by a group) and, more generally, simplicial sets (when it is a more complicated quotient). The starting point of derived algebraic geometry is that its fundamental objects are functors

$$sComm \longrightarrow sSet$$

from the category of simplicial commutative rings to the category of simplicial sets. The main objective of the series of lectures is to explain how the basic notions of algebraic geometry (schemes, algebraic spaces, flat, smooth, and étale morphisms) can be extended to this derived setting, and how this is useful for the study of moduli problems.

We will proceed in two steps. We will first explain how to do half of the job and to allow derived quotients, but not derived tensor products (i.e., considering functors $Comm \longrightarrow sSet$). In other words, we will start to explain formally how the quotient problem (step 2 of the preceding section) can be solved. This will be done by introducing the notions of *stacks* and *algebraic stacks*, which are based on the well-known homotopy theory of simplicial presheaves of Jardine and Joyal. Later on we will explain how to incorporate derived tensor products and simplicial commutative rings to the picture.

# Lecture 2

# Simplicial presheaves as stacks

The purpose of this second lecture is to present the homotopy theory of simplicial presheaves on a Grothendieck site and explain how these are models for stacks. In the next lecture, simplicial presheaves will be used to produce models for (higher) stacks in the context of algebraic geometry and will allow us to define the notion of algebraic $n$-stacks, a far-reaching generalization of the notion of schemes for which all quotients by reasonable equivalence relations exist.

## 2.1  Review of the model category of simplicial presheaves

We let $(C, \tau)$ be a Grothendieck site. Recall that this means that we are given a category $C$ together with a Grothendieck topology $\tau$ on $C$. The Grothendieck topology $\tau$ is the data for any object $X \in C$ of a family $\mathrm{cov}(X)$ of sieves over $X$ (i.e., subfunctors of the representable functor $h_X = \mathrm{Hom}(-, X)$) satisfying the following three conditions:

1. For any $X \in C$, we have $h_X \in \mathrm{cov}(X)$.

2. For any morphism $f \colon Y \longrightarrow X$ in $C$ and any $u \in \mathrm{cov}(X)$, we have

$$f^*(u) = u \times_{h_X} h_Y \in \mathrm{cov}(Y).$$

3. Let $X \in C$, $u \in \mathrm{cov}(X)$, and let $v$ be any sieve on $X$. If for all $Y \in C$ and any $f \in u(Y) \subset \mathrm{Hom}(Y, X)$ we have $f^*(v) \in \mathrm{cov}(Y)$, then $v \in \mathrm{cov}(X)$.

Recall that for such a Grothendieck site we have its associated category of presheaves $Pr(C)$, which by definition is the category of all functors from $C^{\mathrm{op}}$ to the category of sets. The full subcategory of sheaves $Sh(C)$ is defined to be the

I. Moerdijk and B. Toën, *Simplicial Methods for Operads and Algebraic Geometry*, Advanced Courses in Mathematics - CRM Barcelona, DOI 10.1007/978-3-0348-0052-5_10, © Springer Basel AG 2010

subcategory of presheaves $F\colon C^{op} \longrightarrow Set$ such that, for any $X \in C$ and any $u \in \mathrm{cov}(X)$, the natural map

$$F(X) \simeq \mathrm{Hom}_{Pr(C)}(h_X, F) \longrightarrow \mathrm{Hom}_{Pr(C)}(u, F)$$

is bijective. A standard result from sheaf theory states that the inclusion functor

$$i\colon Sh(C) \longhookrightarrow Pr(C)$$

has an exact (i.e., commuting with finite limits) left adjoint

$$a\colon Pr(C) \longrightarrow Sh(C)$$

called the *associated sheaf functor*.

We now let $sPr(C)$ be the category of simplicial objects in $Pr(C)$. We start to endow the category $sPr(C)$ with a levelwise model category structure defined as follows.

**Definition 2.1.1.** Let $f\colon F \longrightarrow F'$ be a morphism in $sPr(C)$.

1.  The morphism $f$ is a *global fibration* if for any $X \in C$ the induced map

    $$F(X) \longrightarrow F'(X)$$

    is a fibration of simplicial sets (for the standard model category structure, i.e., a Kan fibration).

2.  The morphism $f$ is a *global equivalence* if for any $X \in C$ the induced map

    $$F(X) \longrightarrow F'(X)$$

    is an equivalence[1] of simplicial sets (again for the standard model category structure on simplicial sets).

3.  The morphism $f$ is a *global cofibration* if it has the right lifting property with respect to every fibration which is also an equivalence.

It is well known that the above definitions endow the category $sPr(C)$ with a cofibrantly generated model category structure. This model category is moreover proper and cellular (in the sense of [Hi]). This model structure will be referred to as the *global model structure*. There is a small set theory problem here when the category $C$ is not small. This problem can be easily solved by fixing universes and will simply be neglected in the sequel.

We now take into account the Grothendieck topology $\tau$ on $C$ in order to refine the global model structure. This is an important step, since when the quotient of a group action on a scheme exists, the presheaf represented by the quotient scheme is

---

[1] In these notes the expression *equivalence* always refers to *weak equivalence*.

certainly not the quotient presheaf. However, for free actions the sheaf represented by the quotient scheme is the quotient sheaf.

We start by introducing the so-called homotopy sheaves of a simplicial presheaf $F: C^{\mathrm{op}} \longrightarrow sSet$. We define a presheaf

$$\pi_0^{\mathrm{pr}}(F): C^{\mathrm{op}} \longrightarrow Set$$

simply by sending $X \in C$ to $\pi_0(F(X))$. In the same way, for any $X \in C$ and any 0-simplex $s \in F(X)_0$, we define presheaves of groups on $C/X$,

$$\pi_i^{\mathrm{pr}}(F, s): (C/X)^{\mathrm{op}} \longrightarrow Gp$$

sending $f: Y \longrightarrow X$ to $\pi_i(F(Y), f^*(s))$. Here, $F(Y)$ is the simplicial set of values of $F$ over $Y$, $f^*(s) \in F(Y)_0$ is the inverse image of the base point $s$, and finally $\pi_i(F(Y), f^*(s))$ denotes the *correct* homotopy groups of the simplicial set $F(Y)$ based at $f^*(s)$. By *correct* we mean either the simplicial (or combinatorial) homotopy groups of a fibrant model for $F(Y)$, or more easily the topological homotopy groups of the geometric realization $|F(Y)|$.

The associated sheaves to the presheaves $\pi_0^{\mathrm{pr}}(F)$ and $\pi_i^{\mathrm{pr}}(F, s)$ will be denoted by $\pi_0(F)$ and $\pi_i(F, s)$. These are called the *homotopy sheaves* of $F$. They are functorial in $F$.

**Definition 2.1.2.** Let $f: F \longrightarrow F'$ be a morphism of simplicial presheaves.

1. The morphism $f$ is a *local equivalence* if it satisfies the following two conditions:

   (a) The induced morphism $\pi_0(F) \longrightarrow \pi_0(F')$ is an isomorphism of sheaves.

   (b) For any $X \in C$, any $s \in F(X)_0$ and any $i > 0$, the induced morphism $\pi_i(F, s) \longrightarrow \pi_i(F', f(s))$ is an isomorphism of sheaves on $C/X$.

2. The morphism $f$ is a *local cofibration* if it is a global cofibration in the sense of Definition 2.1.1.

3. The morphism $f$ is a *local fibration* if it has the left lifting property with respect to every local cofibration which is also a local equivalence.

For simplicity, we will use the expressions *equivalence, fibration* and *cofibration* in order to refer to *local equivalence, local fibration* and *local cofibration*.

It is also well known that the above definition endows the category $sPr(C)$ with a model category structure, but this is a much harder result than the existence of the global model structure. This result, as well as several small modifications, is due to Joyal (for simplicial sheaves) and Jardine (for simplicial presheaves), and we refer to [Bl, DHI] for recent references. Unless the contrary is specified, we will always assume that the category $sPr(C)$ is endowed with this model category structure, which will be called the *local model structure*.

A nice result proved in [DHI] is the following characterization of fibrant objects in $sPr(C)$ (for the local model structure). Recall first that a *hypercovering* of an object $X \in C$ is the data of a simplicial presheaf $H$ together with a morphism $H \longrightarrow X$ satisfying the following two conditions:

1. For any integer $n$, the presheaf $H_n$ is a disjoint union of representable presheaves.

2. For any $n \geq 0$, the morphism of presheaves (of sets)

$$H_n \simeq \mathrm{Hom}(\Delta^n, H) \longrightarrow \mathrm{Hom}(\partial\Delta^n, H) \times_{\mathrm{Hom}(\partial\Delta^n, X)} \mathrm{Hom}(\Delta^n, X)$$

   induces an epimorphism on the associated sheaves.

Here $\Delta^n$ denotes the simplicial simplex of dimension $n$ as well as the corresponding constant simplicial presheaf. In the same way $\partial\Delta^n$ is the $(n-1)$-skeleton of $\Delta^n$ and is considered here as a constant simplicial presheaf. Finally, Hom denotes here the presheaf of morphisms between two simplicial presheaves. This second condition can equivalently be stated by saying that, for any $Y \in C$ and any commutative square of simplicial sets

$$
\begin{array}{ccc}
\partial\Delta^n & \longrightarrow & H(Y) \\
\downarrow & & \downarrow \\
\Delta^n & \longrightarrow & X(Y) = \mathrm{Hom}(Y, X),
\end{array}
$$

there exists a covering sieve $u \in \mathrm{cov}(Y)$ such that for any $f \colon U \longrightarrow Y$ in the sieve $u$ there exists a morphism $\Delta^n \longrightarrow H(U)$ making the following square commutative:

$$
\begin{array}{ccc}
\partial\Delta^n & \longrightarrow & H(U) \\
\downarrow & \nearrow & \downarrow \\
\Delta^n & \longrightarrow & X(U) = \mathrm{Hom}(U, X).
\end{array}
$$

This property is also called the *local lifting property*, and is the local analog of the lifting property characterizing trivial fibrations of simplicial sets. In particular, it implies that the homotopy sheaves of $H$ and of $X$ coincide, and thus that $H \longrightarrow X$ is a local equivalence. In low dimensions, this local lifting condition says the following:

- $(n = 0)$ The morphism $H_0 \longrightarrow X$ induces an epimorphism on the associated sheaves.

- $(n = 1)$ The morphism $H_1 \longrightarrow H_0 \times_X H_0$ induces an epimorphism on the associated sheaves.

- $(n = 2)$ The morphism $H_2 \longrightarrow H_1 \times_{H_0 \times H_0} (H_1 \times_{H_0} H_1)$ induces an epimorphism on the associated sheaves.

Note that, for $n > 1$, the simplicial set $\partial \Delta^n$ is connected, and thus the morphism $X^{\Delta^n} \longrightarrow X^{\partial \Delta^n}$ is, in this case, an isomorphism. This implies that, for $n > 1$, the condition 2 of being a hypercovering is equivalent to the simpler condition that

$$H_n \longrightarrow H^{\partial \Delta^n}$$

induces an epimorphism on the associated sheaves. Therefore, this condition only depends on $H_n$ for $n > 1$, and not upon $X$ or upon the morphism $H \longrightarrow X$.

For $F \in sPr(C)$, any $X \in C$ and any hypercovering $H$ of $X$, we can define an augmented cosimplicial diagram of simplicial sets

$$F(X) \longrightarrow ([n] \longmapsto F(H_n)).$$

Here, each $H_n$ is a coproduct of representables, say $H_n = \coprod_i H_{n,i}$, and by definition we set $F(H_n) = \prod_i F(H_{n,i})$. (When this product is infinite, it should be taken with some care, by first replacing each $F(H_{n,i})$ by its fibrant model.)

With these notions and notations, it is possible to prove (see [DHI]) that an object $F \in sPr(C)$ is fibrant (for the local model structure) if and only if it satisfies the following two conditions:

1. For any $X \in C$, the simplicial set $F(X)$ is fibrant.

2. For any $X \in C$ and any hypercovering $H \longrightarrow X$, the natural map

$$F(X) \longrightarrow \mathrm{Holim}_{[n] \in \Delta}\, F(H_n)$$

   is an equivalence of simplicial sets.

The above first condition is rather mild and the second condition is of course the important one. It is a homotopy analog of the sheaf condition, in the sense that when $F$ is a presheaf of sets, considered as a simplicial presheaf constant in the simplicial direction, then this second condition for $F$ is equivalent to the fact that $F$ is a sheaf (this is because the homotopy limit is then simply a usual limit in the category of sets and the condition becomes the usual descent condition for sheaves).

**Definition 2.1.3.**  1. An object $F \in sPr(C)$ is called a *stack* if, for any $X \in C$ and any hypercovering $H \longrightarrow X$, the natural map

$$F(X) \longrightarrow \mathrm{Holim}_{[n] \in \Delta}\, F(H_n)$$

is an equivalence of simplicial sets.

2. The homotopy category $\mathrm{Ho}(sPr(C))$ will be called the *homotopy category of stacks* on the site $(C, \tau)$ (or simply the *category of stacks*). Most often objects in $\mathrm{Ho}(sPr(C))$ will simply be called *stacks*. The expressions *morphism of stacks* and *isomorphism of stacks* will refer to morphisms and isomorphisms in $\mathrm{Ho}(sPr(C))$. The set of morphisms of stacks from $F$ to $F'$ will be denoted by $[F, F']$.

The following exercise is to understand that, for a given simplicial presheaf, being a stack and being a sheaf of simplicial sets are two different notions having nothing in common.

**Exercise 2.1.4.**    1. Show that a presheaf of sets $F \colon C^{\mathrm{op}} \longrightarrow Set$, considered as a simplicial presheaf, is a stack if and only if it is a sheaf of sets.

2. Let $G$ be a sheaf of groups on $C$ and consider the simplicial presheaf

$$BG \colon \quad C^{\mathrm{op}} \quad \longrightarrow \quad sSet$$
$$X \quad \longmapsto \quad B(G(X)).$$

Here, if $H$ is any discrete group, $BH$ is its simplicial classifying space, whose set of $n$-simplices is $H^n$ and whose face maps are given by the group structure together with the various projections, and whose degeneracies are given by the generalized diagonal maps. Prove that $BG$ is a sheaf of simplicial sets, but that it is not a stack as soon as there exists an object $X \in C$ with $H^1(X, G) \neq *$.

3. For any $X \in C$, we let $\mathcal{F}(X)$ be the nerve of the groupoid of sheaves of sets over $X$ (its objects are sheaves of sets on the site $C/X$ and its morphisms are isomorphisms between such sheaves). Show how to make $X \longmapsto \mathcal{F}(X)$ into a simplicial presheaf on $C$. Show that $\mathcal{F}$ is a stack which is not a sheaf of simplicial sets in general (e.g., show that the set-valued presheaf of 0-simplices $\mathcal{F}_0$ is not a sheaf on $C$).

In the sequel, we will often use the following terminology and notations.

- For a diagram of stacks $F \longrightarrow H \longleftarrow G$, we denote by $F \times^h_H G$ the corresponding homotopy fiber product of simplicial presheaves (note that this construction is not functorially defined on $\mathrm{Ho}(sPr(C))$ and requires some lift of the diagrams to $sPr(C)$).

- A morphism of stacks $F \longrightarrow F'$ in $\mathrm{Ho}(sPr(C))$ is an *epimorphism* if the induced morphism $\pi_0(F) \longrightarrow \pi_0(F')$ is a sheaf epimorphism.

- A morphism of stacks $F \longrightarrow F'$ is a *monomorphism* if the diagonal morphism $F \longrightarrow F \times^h_{F'} F$ is an isomorphism.

**Exercise 2.1.5.**    1. Show that a morphism of stacks $f \colon F \longrightarrow F'$ is a monomorphism if and only if it satisfies the following two conditions:

   (a) The induced morphism $\pi_0(F) \longrightarrow \pi_0(F')$ is a monomorphism.

   (b) For all $X \in C$ and all $s \in F(X)$, the induced morphisms $\pi_i(F, s) \longrightarrow \pi_i(F', f(s))$ are isomorphisms for all $i > 0$.

2. Show that a morphism of stacks $F \longrightarrow F'$ is an epimorphism (resp. a monomorphism) if and only if, for any $X \in C$ and any morphism $X \longrightarrow F'$ in $\mathrm{Ho}(sPr(C))$, the induced projection $F \times^h_{F'} X \longrightarrow X$ is an epimorphism (resp. a monomorphism).

## 2.2 Basic examples

The most fundamental example of a stack is the *stack of stacks*, whose existence expresses that *stacks can be defined locally and glued*. This example of a stack is important for conceptual reasons, but also because it can be used to construct many examples of other stacks. Its precise definition goes as follows. For $X \in C$, we consider the category $sPr^W(C/X)$ whose objects are simplicial presheaves on $C/X$ and whose morphisms are the local equivalences. For a morphism $f : Y \longrightarrow X$ in $C$, we have a base-change functor $sPr^W(C/X) \longrightarrow sPr^W(C/Y)$ which makes $X \longmapsto sPr^W(C/X)$ into a presheaf of categories. Taking the nerve of all the categories $sPr^W(C/X)$, we obtain a simplicial presheaf

$$\begin{aligned} S: \quad C^{\mathrm{op}} &\longrightarrow sSet \\ X &\longmapsto N(sPr^W(C/X)). \end{aligned}$$

**Theorem 2.2.1.** *The simplicial presheaf $S$ defined above is a stack. It is called the stack of stacks.*

*Sketch of a proof:* There are several different ways to prove the above theorem, unfortunately none of them being really easy. We sketch here the main steps for one of them, but a complete and detailed proof would be much too long for these notes, as well as not very instructive.

**Step 1:** We first extend the functor $S: C^{\mathrm{op}} \longrightarrow sSet$ to a functor

$$S': sPr(C)^{\mathrm{op}} \longrightarrow sSet.$$

This will cause some set-theoretic troubles because $sPr(C)$ is a *non-small* category. This issue can be solved in at least two different ways: either by using universes or by choosing a large enough bound on the cardinalities of the presheaves that we want to consider. For $F \in sPr(C)^{\mathrm{op}}$, we consider the category $Fib^W/F$ whose objects are fibrations $F' \longrightarrow F$ in $sPr(C)$ and whose morphisms are local equivalences in $sPr(C)/F$. Each $F' \longrightarrow F$ determines a base-change functor

$$F' \times_F - : Fib^W/F \longrightarrow Fib^W/F'.$$

This does not define a presheaf of categories on $sPr(C)$, but only a lax functor. However, any lax functor is equivalent to a strict functor by a natural construction called rectification (see for example [Hol, §3.3]). We will therefore proceed as if $F \longmapsto Fib^W/F$ were a genuine presheaf of categories. Taking the nerves of all the categories $Fib^W/F$ provides a simplicial presheaf

$$\begin{aligned} S': \quad sPr(C)^{\mathrm{op}} &\longrightarrow sSet \\ F &\longmapsto N(Fib^W/F). \end{aligned}$$

This simplicial presheaf, restricted to $C \hookrightarrow sPr(C)$, is naturally equivalent to $S$, as its value on $X \in C$ is the nerve of the category of equivalences between fibrant objects in $sPr(C)/X \simeq sPr(C/X)$. This nerve is itself equivalent to the nerve of local equivalences between all simplicial objects in $sPr(C/X)$, as the fibrant replacement functor gives an inverse up to homotopy of the natural inclusion.

The conclusion of this first step is that $S$ possesses, up to a natural equivalence, an extension $S'$ as a presheaf on $sPr(C)$.

**Step 2:** The functor $S' \colon sPr(C)^{\mathrm{op}} \longrightarrow sSet$ sends local equivalences to equivalences, and homotopy colimits in $sPr(C)$ to homotopy limits in $sSet$. Indeed, the fact that $S'$ preserves equivalences follows formally from the fact that the model category $sPr(C)$ is right proper. That it sends homotopy colimits to homotopy limits is more subtle. First of all, any homotopy colimit can be obtained by a succession of homotopy pushouts and homotopy disjoint unions. Therefore, in order to prove that $S'$ sends homotopy colimits to homotopy limits it is enough to prove the following two statements:

1. For any family of objects $\{F_i\}_{i \in I}$ in $sPr(C)$, the natural morphism

$$S' \left( \coprod_i F_i \right) \longrightarrow \prod_i S'(F_i)$$

    is a weak equivalence.

2. For any homotopy pushout diagram in $sPr(C)$

$$\begin{array}{ccc} F_0 & \longrightarrow & F_1 \\ \downarrow & & \downarrow \\ F_2 & \longrightarrow & F, \end{array}$$

    the induced diagram

$$\begin{array}{ccc} S'(F) & \longrightarrow & S'(F_2) \\ \downarrow & & \downarrow \\ S'(F_1) & \longrightarrow & S'(F_0) \end{array}$$

    is homotopy cartesian in $sSet$.

Statement 1 follows from the fact that the model category $sPr(C)/(\coprod_i F_i)$ is the product of the model categories $sPr(C)/F_i$ (there is a small issue with infinite products that we do not mention here). Statement 2 is the key of the proof of the theorem and is the hardest point. It can be deduced from [Re, Theorem 1.4] as follows. We assume that we have a diagram

$$F_1 \xleftarrow{\ \ i\ \ } F_0 \xrightarrow{\ \ j\ \ } F_2,$$

with $i$ and $j$ cofibrations between cofibrant objects (requiring one of the two morphisms to be a cofibration would be enough here). We let $F$ be the pushout of this diagram in $sPr(C)$, which is therefore also a homotopy pushout. We have a diagram of Quillen adjunctions obtained by base change (we write here the right adjoints):

$$sPr(C)/F \longrightarrow sPr(C)/F_1$$
$$\downarrow \qquad\qquad \downarrow$$
$$sPr(C)/F_2 \longrightarrow sPr(C)/F_0.$$

Because of [Re, Theorem 1.4], this diagram of model categories satisfies the (opposite) conditions of [To2, Lemma 4.2], which ensures that the corresponding diagram of simplicial sets obtained by taking the nerve of the categories of equivalences between fibrant objects

$$Fib^W/F \longrightarrow Fib^W/F_1$$
$$\downarrow \qquad\qquad \downarrow$$
$$Fib^W/F_2 \longrightarrow Fib^W/F_0$$

is homotopy cartesian.

**Step 3:** We can now conclude from steps 1 and 2 that $\mathcal{S}$ is a stack. Indeed, let $H \longrightarrow X$ be a hypercovering. The morphism

$$\mathcal{S}(X) \longrightarrow \text{Holim}_n \ \mathcal{S}(H_n)$$

is equivalent to

$$\mathcal{S}'(X) \longrightarrow \text{Holim}_n \ \mathcal{S}'(H_n).$$

But, as $\mathcal{S}'$ converts homotopy colimits to homotopy limits, we see that this last morphism is an equivalence because the morphism

$$H \simeq \text{Hocolim}_n \ H_n \longrightarrow X$$

is a local equivalence. $\qquad\qquad\qquad\qquad\qquad\qquad\qquad\qquad\qquad\qquad\qquad\square$

To finish this section, we present some basic and general examples of stacks. These are very general examples and we will see more specific examples in the context of algebraic geometry in the next lectures.

**Sheaves:** We start by noticing that there is a full embedding

$$Sh(C) \longrightarrow \text{Ho}(sPr(C))$$

from the category of sheaves (of sets) to the homotopy category of stacks, simply by considering a sheaf of sets as a simplicial presheaf (constant in the simplicial direction). This inclusion functor has a left adjoint, which sends a simplicial presheaf $F$ to the sheaf $\pi_0(F)$. This will allow us to consider any sheaf as a stack, and in the sequel we will do this implicitly. In this way, the category of stacks $\text{Ho}(sPr(C))$ is an extension of the category of sheaves. Moreover, any object in $\text{Ho}(sPr(C))$ is isomorphic to a homotopy colimit of sheaves (this is because any simplicial set $X$ is naturally equivalent to the homotopy colimit of the diagram $[n] \longmapsto X_n$), which shows that stacks are obtained from sheaves by taking derived quotients.

**Classifying stacks:** Let $G$ be a group object in $sPr(C)$, that is, a presheaf of simplicial groups. From it we construct a simplicial presheaf $BG$ by applying levelwise the classifying space construction. More explicitly, $BG$ is the simplicial presheaf whose presheaf of $n$-simplices is $(G_n)^n$, and whose faces and degeneracies are defined using the composition and units in $G$ as well as the faces and degeneracies of the underlying simplicial set of $G$. The simplicial presheaf $BG$ has a natural global point $*$, and by construction we have

$$\pi_i(BG, *) \simeq \pi_{i-1}(G, e).$$

When $G$ is abelian, the simplicial presheaf $BG$ is again an abelian group object in $sPr(C)$, and the construction can then be iterated.

When $A$ is a sheaf of abelian groups on $C$, we let

$$K(A, n) = \underbrace{B \ldots B}_{n \text{ times}}(A).$$

By construction, $K(A, n)$ is a pointed simplicial presheaf such that

$$\pi_i(K(A, n), *) = 0 \text{ if } i \neq n, \qquad \pi_n(K(A, n), *) \simeq A,$$

and this characterizes $K(A, n)$ uniquely up to an isomorphism in $\text{Ho}(sPr(C))$.

**Exercise 2.2.2.**   1. Let $X \in C$ and let $H \longrightarrow X$ be a hypercovering. Show that there exist natural isomorphisms

$$\pi_i(\text{Holim}_{[n] \in \Delta} K(A, n)(H_n)) \simeq \check{H}^{n-i}(H/X, A),$$

where the right-hand side is Čech cohomology of $X$ with coefficients in $A$ with respect to the hypercovering $H$ (see [Ar-Ma]).

2. Deduce from part 1 that the simplicial presheaf $K(A, n)$ is a stack if and only if $A$ is a locally acyclic sheaf (i.e., for any $X \in C$ we have $H^i(X, A) = 0$ for $i > 0$).

3. Use part 2 and induction on $n$ to prove that for any $X \in C$ there exist natural isomorphisms

$$[X, K(A, n)] \simeq H^n(X, A),$$

where the left-hand side is the set of morphisms in $\text{Ho}(sPr(C))$ and the right-hand side denotes sheaf cohomology.

**Truncations and $n$-stacks:** A stack $F \in \text{Ho}(sPr(C))$ is *$n$-truncated*, or an *$n$-stack*, if for any $X \in C$ and any $s \in F(X)_0$ we have $\pi_i(F, s) = 0$ for all $i > n$. The full subcategory of $n$-stacks will be denoted by $\text{Ho}(sPr_{\leq n}(C))$. We note that $\text{Ho}(sPr_{\leq 0}(C))$ is the essential image of the inclusion morphism

$$Sh(C) \longrightarrow \text{Ho}(sPr(C)),$$

and thus that there is an equivalence of categories $Sh(C) \simeq \text{Ho}(sPr_{\leq 0}(C))$.

The inclusion functor $\text{Ho}(sPr_{\leq n}(C)) \hookrightarrow \text{Ho}(sPr(C))$ admits a left adjoint

$$t_{\leq n} \colon \text{Ho}(sPr(C)) \longrightarrow \text{Ho}(sPr_{\leq n}(C))$$

called the *truncation functor*. We have $t_{\leq 0} \simeq \pi_0$, and in general $t_{\leq n}$ is obtained by applying levelwise the usual truncation functor for simplicial sets. Another possible understanding of this situation is by introducing the left Bousfield localization (in the sense of [Hi]) of the model category $sPr(C)$ by inverting all the morphisms $\partial \Delta^{n+2} \times X \longrightarrow X$, for all $X \in C$. The fibrant objects for this localized model structure are precisely the $n$-truncated fibrant simplicial presheaves, and its homotopy category can be naturally identified with $\text{Ho}(sPr_{\leq n}(C))$. The functor $t_{\leq n}$ is then the localization functor for this left Bousfield localization.

For any stack $F$, there exists a tower of stacks

$$F \longrightarrow \cdots \longrightarrow t_{\leq n}(F) \longrightarrow t_{\leq n-1}(F) \longrightarrow \cdots \longrightarrow t_{\leq 0}(F) = \pi_0(F),$$

called the *Postnikov tower* for $F$. A new feature here is that this tower does not converge in general, or in other words the natural morphism

$$F \longrightarrow \text{Holim}_n \, t_{\leq n}(F)$$

is not an equivalence in general. It is the case under some rather strong boundedness conditions on the cohomological dimension of the sheaves of groups $\pi_i(F)$.

**Exercise 2.2.3.** Suppose that there exists an integer $d$ such that, for any $X \in C$ and any sheaf of abelian groups $A$ on $C/X$, we have $H^i(X, A) = 0$ for all $i > d$. Prove that, for any stack $F$, the natural morphism

$$F \longrightarrow \text{Holim}_n \, t_{\leq n}(F)$$

is an isomorphism in $\text{Ho}(sPr(C))$.

**Internal Hom:** An important property of the category $\text{Ho}(sPr(C))$ is that it admits internal Homs (i.e., is cartesian closed). One way to see this is to use the injective model structure on $sPr(C)$, originally introduced in [Ja]. In order to distinguish this model structure from the projective model structure that we are using in these notes, we denote by $sPr_{\text{inj}}(C)$ the category of simplicial presheaves endowed with the injective model structure. Its equivalences are the local equivalences of Definition 2.1.2, and its cofibrations are the monomorphisms of simplicial presheaves. The nice property of the model category $sPr_{\text{inj}}(C)$ is that it becomes a monoidal model category in the sense of [Ho] when endowed with the monoidal structure given by the direct product. A formal consequence of this is that $\mathrm{Ho}(sPr_{\mathrm{inj}}(C)) = \mathrm{Ho}(sPr(C))$ is cartesian closed (see for instance [Ho, Theorem 4.3.2]).

Explicitly, if $F$ and $F'$ are two stacks, we define a simplicial presheaf

$$\mathbb{R}\mathcal{H}om(F, F') \colon C^{\mathrm{op}} \longrightarrow sSet$$

by

$$\mathbb{R}\mathcal{H}om(F, F')(X) = \underline{\mathrm{Hom}}(X \times F, R(F')),$$

where $\underline{\mathrm{Hom}}$ denotes the natural simplicial enrichment of the category $sPr(C)$ and $R(F')$ is a fibrant model for $F'$ as an object in $sPr_{\mathrm{inj}}(C)$. When the object $\mathbb{R}\mathcal{H}om(F, F')$ is considered in $\mathrm{Ho}(sPr(C))$, it is possible to show that we have functorial isomorphisms

$$[F'', \mathbb{R}\mathcal{H}om(F, F')] \simeq [F'' \times F, F']$$

for any $F'' \in \mathrm{Ho}(sPr(C))$. The stack $\mathbb{R}\mathcal{H}om(F, F')$ is called the *stack of morphisms* from $F$ to $F'$.

**Exercise 2.2.4.** Assume that $C$ possesses finite products. Prove that the model category $sPr(C)$, as usual with its projective local model structure, is a monoidal model category for the monoidal structure given by the direct product (for this, use e.g. [Ho, Corollary 4.2.5] and the explicit generating cofibrations $A \times X \longrightarrow B \times X$, where $A \longrightarrow B$ is a cofibration in $sSet$).

**Substacks defined by local conditions:** Let $F$ be a stack on $C$, and $F_0$ its presheaf of 0-dimensional simplices. A *condition on objects* on $F$ is, by definition, a subpresheaf $G_0 \subset F_0$. Such a condition is *saturated* if there exists a pullback square

$$
\begin{array}{ccc}
G_0 & \longrightarrow & F_0 \\
\downarrow & & \downarrow \\
E & \longrightarrow & \pi_0^{\mathrm{pr}}(F).
\end{array}
$$

Equivalently, the condition is saturated if, for any $X \in C$, the subset $G_0(X) \subset F_0(X)$ is a union of connected components. Finally, we say that a saturated condition on $F$ is *local* if, for any $X \in C$ and any $x \in F_0(X)$, we have

$$(x \in G_0(X)) \Longleftrightarrow (\exists u \in \mathrm{cov}(X) \text{ such that } f^*(x) \in G_0(U) \; \forall f \colon U \longrightarrow X \text{ in } u).$$

Let $G_0 \subset F_0$ be a saturated local condition on $F$. We define a sub-simplicial presheaf $G \subset F$ as follows. For $[n] \in \Delta$, $G(X)_n$ is the subset of $F(X)_n$ consisting of all $n$-dimensional simplices $\alpha$ such that all 0-dimensional faces of $\alpha$ belong to $G_0(X)$. As the condition is saturated, $G(X)$ is a union of connected components of $F(X)$. Moreover, since the condition is local and $F$ is a stack, it is easily seen that $G$ is itself a stack. The stack $G$ is called the *substack of $F$ defined by the condition $G_0$*.

**Twisted forms:** Let $F \in \mathrm{Ho}(sPr(C))$ be a stack. We consider the following condition $G_F$ on the stack of stacks $\mathcal{S}$. For $X \in C$, the set $\mathcal{S}(X)_0$ is by definition the set of simplicial presheaves on $C/X$. We let $G_F(X) \subset \mathcal{S}(X)_0$ be the subset corresponding to all simplicial presheaves $F'$ such that there exists a covering sieve $u \in \mathrm{cov}(x)$ such that, for all $U \longrightarrow X$ in $u$, the restrictions of $F$ and $F'$ are isomorphic in $\mathrm{Ho}(sPr(C/U))$. This condition is a saturated and local condition on $\mathcal{S}$, and therefore defines, as explained in our previous example, a substack $\mathcal{S}_F \subset \mathcal{S}$. The stack $\mathcal{S}_F$ is called the *stack of twisted forms* of $F$.

**Exercise 2.2.5.** Let $F$ and $G$ be two stacks. Prove that there exists a natural bijection between $[G, \mathcal{S}_F]$ and the subset of isomorphism classes of objects in $\mathrm{Ho}(sPr(C)/G)$ consisting of all objects $G' \longrightarrow G$ satisfying the following condition: There is a family of objects $\{X_i\}$ of $C$ and an epimorphism $\coprod_i X_i \longrightarrow G$ such that, for any $i$, the stack $G' \times_G^h X_i$ is isomorphic in $\mathrm{Ho}(sPr(C/X_i))$ to the restriction of $F$.

**More about twisted forms:** Let $F$ be a given stack, for which we want to understand better the stack of twisted forms $\mathcal{S}_F$. We consider the presheaf of simplicial monoids

$$\mathbb{R}\mathcal{E}nd(F)\colon X \longmapsto \underline{\mathrm{Hom}}(X \times R(F), R(F)),$$

where $R(F)$ is an injective fibrant model for $F$. The monoid structure on this presheaf is induced by composing endomorphisms. We define another presheaf of simplicial monoids by the following homotopy pullback square:

$$
\begin{array}{ccc}
\mathbb{R}\underline{Aut}(F) & \longrightarrow & \mathbb{R}\mathcal{E}nd(F) \\
\downarrow & & \downarrow \\
\pi_0(\mathbb{R}\mathcal{E}nd(F))^{\mathrm{inv}} & \longrightarrow & \pi_0(\mathbb{R}\mathcal{E}nd(F)),
\end{array}
$$

where $\pi_0(\mathbb{R}\mathcal{E}nd(F))^{\mathrm{inv}}$ denotes the subsheaf of invertible elements in the sheaf of monoids $\pi_0(\mathbb{R}\mathcal{E}nd(F))$.

The stack $\mathbb{R}\underline{Aut}(F)$ is called the *stack of auto-equivalences* of $F$. It is represented by a presheaf in simplicial monoids for which all elements are invertible up to homotopy. Even if this is not, strictly speaking, a presheaf of simplicial groups, we can apply the classifying space construction to get a new stack $B\mathbb{R}\underline{Aut}(F)$. There exists a natural morphism of stacks

$$B\mathbb{R}\underline{Aut}(F) \longrightarrow \mathcal{S}$$

constructed as follows. The simplicial monoid $\underline{Aut}(F)$ acts on the simplicial presheaf $R(F)$ in an obvious way. We form the Borel construction for this action to get a new simplicial presheaf $[F/\underline{Aut}(F)]$, which is, by definition, the homotopy colimit of the standard simplicial object

$$([n] \longmapsto \underline{Aut}(F)^n \times F).$$

There exists a natural projection

$$[F/\underline{Aut}(F)] \longrightarrow [*/\underline{Aut}(F)] = B/\underline{Aut}(F),$$

giving an object in $\mathrm{Ho}(sPr(C)/(B/\underline{Aut}(F)))$. By Exercise 2.2.5, this object corresponds to a well-defined morphism in $\mathrm{Ho}(sPr(C))$,

$$B\mathbb{R}\underline{Aut}(F) \longrightarrow \mathcal{S}_F \subset \mathcal{S}.$$

Moreover, [HAGII, Proposition A.0.6] implies that the morphism

$$B\mathbb{R}\underline{Aut}(F) \longrightarrow \mathcal{S}_F$$

is in fact an isomorphism.

An important example is when $F = K(A, n)$, for $A$ a sheaf of abelian groups, as twisted forms of $F$ are sometimes referred to as *n-gerbes with coefficients in A*. It can be shown that the monoid $\mathbb{R}\underline{Aut}(F)$ is the semi-direct product of $K(A, n)$ by the sheaf of groups $\mathrm{aut}(A)$. In other words, we have

$$\mathcal{S}_{K(A,n)} \simeq [K(A+1, n)/\mathrm{aut}(A)],$$

or, in other words, we have a split fibration sequence

$$K(A, n+1) \longrightarrow \mathcal{S}_{K(A,n)} \longrightarrow \mathrm{Baut}(A).$$

As a consequence, we see that the set of equivalence classes of $n$-gerbes on $X$ with coefficients in $A$ is in bijection with the set of pairs $(\rho, \alpha)$, where $\rho \in H^1(X, \mathrm{aut}(A))$ and $\alpha \in H^{n+1}(X, A_\rho)$, where $A_\rho$ is the twisted form of $A$ determined by $\rho$.

Another important application is the description of the inductive construction of the Postnikov tower of a stack $F$. We first assume that $F$ is simply connected and connected (i.e., that the sheaves $\pi_0(F)$ and $\pi_1(F)$ are trivial on $C$). This implies that we have globally defined sheaves $\pi_n(F)$ on $C$. The natural projection in the Postnikov tower

$$t_{\leq n+1}(F) \longrightarrow t_{\leq n}(F)$$

produces an object in $\mathrm{Ho}(sPr(C))/t_{\leq n}(F)$, which, by Exercise 2.2.5, produces a    .
morphism of stacks

$$t_{\leq n}(F) \longrightarrow \mathcal{S}_{K(\pi_n(F),n)},$$

which is the $n$-th Postnikov invariant of $F$. It determines, in particular, a class $k_n \in H^{n+1}(t_{\leq n}(F), \pi_n(F))$ which completely determines the object $t_{\leq n+1}(F)$. For a general $F$, possibly non-connected and non-simply connected, it is possible, by changing the base site $C$, to reduce to the connected and simply connected case.

**Exercise 2.2.6.** Let $F$ be any stack and let $\Pi_1(F)$ be its associate presheaf of fundamental groupoids, sending $X$ to $\Pi_1(F(X))$. We let $p\colon D = \int_C \Pi_1(F) \longrightarrow C$ be the Grothendieck construction of the functor $\Pi_1(F)$, endowed with the topology induced from the one on $C$. Recall that objects in $D$ are pairs $(X, x)$ consisting of $X \in C$ and $x \in \Pi_1(F(X))$, and morphisms $(X, x) \longrightarrow (Y, y)$ consist of pairs $(u, f)$, where $u\colon X \longrightarrow Y$ is in $C$ and $f\colon u^*(x) \longrightarrow y$ is in $\Pi_1(F(Y))$.

1. Show that the functor

$$D \longrightarrow \mathrm{Ho}(sPr(C)/t_{\leq 1}(F)),$$

   sending $(X, x)$ to $x\colon X \longrightarrow F$, extends to an equivalence

$$\mathrm{Ho}(sPr(D)) \simeq \mathrm{Ho}(sPr(C)/t_{\leq 1}(F)).$$

2. Show that, under this equivalence, the image of $F \longrightarrow t_{\leq 1}(F)$ is a connected and simply connected object in $\mathrm{Ho}(sPr(D))$.

3. Deduce the existence of Postnikov invariants $k_n \in H^{n+1}(t_{\leq n}(F), \pi_n(F))$ for $F$, where the cohomology group now means cohomology with coefficients in a sheaf $\pi_n(F)$ living on $t_{\leq 1}(F)$.

# Lecture 3

# Algebraic stacks

In the previous lecture we introduced the notion of stacks over some site. We will now consider the more specific case of stacks over the étale site of affine schemes and introduce an important class of stacks called *algebraic stacks*. These are generalizations of schemes and algebraic spaces for which quotients by smooth actions always exist.

Throughout this lecture we will consider the category *Comm* of commutative rings and set $Aff = Comm^{op}$. For $A \in Comm$, we denote by $\operatorname{Spec} A$ the corresponding object in *Aff* (therefore "Spec" is a formal notation here). We endow *Aff* with the étale topology defined as follows. Recall that a morphism of commutative rings $A \longrightarrow B$ is *étale* if it satisfies the following three conditions:

1. $B$ is flat as an $A$-module.

2. $B$ is finitely presented as a commutative $A$-algebra; that is, of the form $A[T_1, \ldots, T_n]/(P_1, \ldots, P_r)$.

3. $B$ is flat as a $B \otimes_A B$-module.

There exist several equivalent characterizations of étale morphisms (see e.g. [SGA1]); for instance, the third condition can be equivalently replaced by the condition $\Omega^1_{B/A} = 0$, where $\Omega^1_{B/A}$ is the $B$-module of relative Kähler derivations (corepresenting the functor sending a $B$-module $M$ to the set of $A$-linear derivations on $B$ with coefficients in $M$). Étale morphisms are stable under base change and composition in *Aff*, i.e., by cobase change and composition in *Comm*. Geometrically, an étale morphism $A \longrightarrow B$ should be thought of as a "local isomorphism" of schemes $\operatorname{Spec} B \longrightarrow \operatorname{Spec} A$, though here *local* should not be understood in the sense of the Zariski topology.

Now, a family of morphisms $\{A \longrightarrow A_i\}_{i \in I}$ is an étale covering if each morphism $A \longrightarrow A_i$ is étale and if the family of base-change functors

$$- \otimes_A A_i \colon A\text{-}Mod \longrightarrow A_i\text{-}Mod$$

is conservative. This defines a topology on *Aff* by defining that a sieve on $\operatorname{Spec} A$ is a covering sieve if it is generated by an étale covering family.

Finally, a morphism $\operatorname{Spec} B \longrightarrow \operatorname{Spec} A$ is a Zariski open immersion if it is étale and a monomorphism (this is equivalent to imposing that the natural morphism $B \otimes_A B \longrightarrow B$ is an isomorphism, or equivalently that the forgetful functor $B\text{-}Mod \longrightarrow A\text{-}Mod$ is fully faithful).

## 3.1   Schemes and algebraic $n$-stacks

We start by the definition of schemes and then define algebraic $n$-stacks as certain succesive quotients of schemes.

For $\operatorname{Spec} A \in Aff$, we can consider the presheaf represented by $\operatorname{Spec} A$,

$$\operatorname{Spec} A \colon Aff^{\mathrm{op}} = Comm \longrightarrow Set,$$

by setting $(\operatorname{Spec} A)(B) = \operatorname{Hom}(A, B)$. A standard result of commutative algebra (faithfully flat descent) states that the presheaf $\operatorname{Spec} A$ is always a sheaf. We thus consider $\operatorname{Spec} A$ as a stack and as an object in $\operatorname{Ho}(sPr(Aff))$. This defines a fully faithful functor

$$Aff \longrightarrow \operatorname{Ho}(sPr(Aff)).$$

Any object in $\operatorname{Ho}(sPr(Aff))$ isomorphic to a sheaf of the form $\operatorname{Spec} A$ will be called an *affine scheme*. The full subcategory of $\operatorname{Ho}(sPr(Aff))$ consisting of affine schemes is equivalent to $Aff = Comm^{\mathrm{op}}$, and these two categories will be implicitly identified.

**Definition 3.1.1.**    1. Let $\operatorname{Spec} A$ be an affine scheme, $F$ a stack and $i \colon F \longrightarrow \operatorname{Spec} A$ a morphism. We say that $i$ is a *Zariski open immersion* (or simply an *open immersion*) if it satisfies the following two conditions:

   (a) The stack $F$ is a sheaf (i.e., 0-truncated) and the morphism $i$ is a monomorphism of sheaves.

   (b) There exists a family of Zariski open immersions $\{A \longrightarrow A_i\}_i$ such that $F$ is the image of the morphism of sheaves

$$\coprod_i \operatorname{Spec} A_i \longrightarrow \operatorname{Spec} A.$$

2. A morphism of stacks $F \longrightarrow F'$ is a *Zariski open immersion* (or simply an *open immersion*) if, for any affine scheme $\operatorname{Spec} A$ and any morphism $\operatorname{Spec} A \longrightarrow F'$, the induced morphism

$$F \times^h_{F'} \operatorname{Spec} A \longrightarrow \operatorname{Spec} A$$

is a Zariski open immersion in the above sense.

3. A stack $F$ is a *scheme* if there exists a family of affine schemes $\{\operatorname{Spec} A_i\}_i$ and Zariski open immersions $\operatorname{Spec} A_i \longrightarrow F$ such that the induced morphism of sheaves

$$\coprod_i \operatorname{Spec} A_i \longrightarrow F$$

is an epimorphism. Such a family of morphisms $\{\operatorname{Spec} A_i \longrightarrow F\}$ will be called a *Zariski atlas* for $F$.

**Exercise 3.1.2.**    1. Show that any Zariski open immersion $F \longrightarrow F'$ is a monomorphism of stacks.

2. Deduce from this fact that a scheme $F$ is always 0-truncated, and thus equivalent to a sheaf.

We now pass to the definition of algebraic stacks. These are stacks obtained by gluing schemes along smooth quotients, and we first need to recall the notion of smooth morphisms of schemes.

Recall that a morphism of commutative rings $A \longrightarrow B$ is *smooth* if it is flat of finite presentation and if moreover $B$ is of finite Tor dimension as a $B \otimes_A B$-module. Smooth morphisms are the algebraic analog of submersions, and there exist equivalent definitions making this analogy more clear (see [SGA1]). Smooth morphisms are stable under composition and base change in *Aff*. The notion of smooth morphisms can be extended to a notion for all schemes in the following way. We say that a morphism of schemes $X \longrightarrow Y$ is *smooth* if there exist Zariski atlases $\{\operatorname{Spec} A_i \longrightarrow X\}$ and $\{\operatorname{Spec} A_j \longrightarrow Y\}$ together with commutative squares

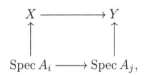

with $\operatorname{Spec} A_i \longrightarrow \operatorname{Spec} A_j$ a smooth morphism —here $j$ depends on $i$. Again, smooth morphisms of schemes are stable under composition and base change.

We are now ready to define the notion of algebraic stack. The definition is by induction on an algebraicity index $n$ representing the number of successive smooth quotients we take. This index will be forgotten after the definition is achieved.

**Definition 3.1.3.**    1. A stack $F$ is *0-algebraic* if it is a scheme.

2. A morphism of stacks $F \longrightarrow F'$ is *0-algebraic* (or *0-representable*) if, for any scheme $X$ and any morphism $X \longrightarrow F'$, the stack $F \times^h_{F'} X$ is 0-algebraic (i.e., a scheme).

3. A 0-algebraic morphism of stacks $F \longrightarrow F'$ is smooth if, for any scheme $X$ and any morphism $X \longrightarrow F'$, the morphism of schemes $F \times^h_{F'} X \longrightarrow X$ is smooth.

4. We now let $n > 0$, and assume that the notions of $(n-1)$-algebraic stack, $(n-1)$-algebraic morphism and smooth $(n-1)$-algebraic morphism have been defined.

   (a) A stack $F$ is *n-algebraic* if there exists a scheme $X$ together with a smooth $(n-1)$-algebraic morphism $X \longrightarrow F$ which is an epimorphism. Such a morphism $X \longrightarrow F$ is called a *smooth n-atlas* for $F$.

   (b) A morphism of stacks $F \longrightarrow F'$ is *n-algebraic* (or *n-representable*) if, for any scheme $X$ and any morphism $X \longrightarrow F'$, the stack $F \times_{F'} X$ is $n$-algebraic.

   (c) An $n$-algebraic morphism of stacks $F \longrightarrow F'$ is smooth if, for any scheme $X$ and any morphism $X \longrightarrow F'$, there exists a smooth $n$-atlas $Y \longrightarrow F \times^h_{F'} X$ such that each morphism $Y \longrightarrow X$ is a smooth morphism of schemes.

5. An *algebraic stack* is a stack which is $n$-algebraic for some integer $n$. An *algebraic n-stack* is an algebraic stack which is also an $n$-stack. An *algebraic space* is an algebraic 0-stack.

6. A morphism of stacks $F \longrightarrow F'$ is *algebraic* (or *representable*) if it is $n$-algebraic for some $n$.

7. A morphism of stacks $F \longrightarrow F'$ is *smooth* if it is $n$-algebraic and smooth for some integer $n$.

Long, but formal arguments show that algebraic stacks satisfy the following properties:

- Algebraic stacks are stable under finite homotopy limits (i.e., by homotopy pullbacks).

- Algebraic stacks are stable under disjoint union.

- Algebraic morphisms of stacks are stable under composition and base change.

- Algebraic stacks are stable under smooth quotients. Thus, if $F \longrightarrow F'$ is a smooth epimorphism of stacks, then $F'$ is algebraic if and only if $F$ is so.

**Exercise 3.1.4.** Let $F$ be an algebraic $n$-stack, $X \in Aff$, and $x \colon X \longrightarrow F$ a morphism of stacks. Show that the sheaf $\pi_n(F, x)$ is representable by an algebraic space, locally of finite type over $X$.

The standard finiteness properties of schemes can be extended to algebraic stacks in the following way:

- An algebraic morphism $F \longrightarrow F'$ is *locally of finite presentation* if, for any scheme $X$ and any morphism $X \longrightarrow F'$, there exists a smooth atlas $Y \longrightarrow F \times^h_{F'} X$ such that the induced morphism $Y \longrightarrow X$ is locally of finite presentation.

- An algebraic morphism $F \longrightarrow F'$ is *quasi-compact* if, for any affine scheme $X$ and any morphism $X \longrightarrow F'$, there exists a smooth atlas $Y \longrightarrow F \times^h_{F'} X$ with $Y$ an affine scheme.

- An algebraic stack $F$ is *strongly quasi-compact* if, for all $n$, the induced morphism

$$F \longrightarrow \mathbb{R}\underline{\mathcal{H}om}(\partial\Delta^n, F)$$

  is quasi-compact.

- An algebraic stack morphism $F \longrightarrow F'$ is *strongly of finite presentation* if, for any affine scheme $X$ and any morphism $X \longrightarrow F'$, the stack $F \times^h_{F'} X$ is locally of finite presentation and strongly quasi-compact.

Note that, when $n = 0$, we have $\mathbb{R}\underline{\mathcal{H}om}(\partial\Delta^n, F) \simeq F \times F$, and the condition of strongly quasi-compactness implies in particular that the diagonal morphism $F \longrightarrow F \times F$ is quasi-compact. In general, being strongly quasi-compact involves quasi-compactness conditions for all the "higher diagonals".

**Exercise 3.1.5.** Let $X$ be an affine scheme and $G$ be a sheaf of groups on $Aff/X$. We form the classifying stack $K(G,1) \in \mathrm{Ho}(sPr(Aff)/X)$, and consider it in $\mathrm{Ho}(sPr(Aff))$.

1. Show that, if $K(G,1)$ is an algebraic stack, then $G$ is represented by an algebraic space locally of finite type.

2. Conversely, if $G$ is representable by an algebraic space which is smooth over $X$, then $K(G,1)$ is an algebraic stack.

3. Assume that $K(G,1)$ is algebraic. Show that $K(G,1)$ is quasi-compact. Show that $K(G,1)$ is strongly quasi-compact if and only if $G$ is quasi-compact.

## 3.2 Some examples

**Classifying stacks:** Suppose that $G$ is a sheaf of groups over some affine scheme $X$, and assume that $G$ is an algebraic space, flat and of finite presentation over $X$. We can form $K(G,1) \in \mathrm{Ho}(sPr(Aff))$, the classifying stack of the group $G$, as explained in §2.2. The stack $K(G,1)$ is however not exactly the right object to consider, at least when $G$ is not smooth over $X$. Indeed, for $Y$ an affine scheme over $X$, $[Y, K(G,1)]$ classifies $G$-torsors over $Y$ which are locally trivial for the étale topology on $Y$. This is a rather unnatural condition, as there exist $G$-torsors, locally trivial for the flat topology on $Y$, which are not étale locally trivial (for instance, when $X = \mathrm{Spec}\, k$ is a perfect field of characteristic $p$, the Frobenius map $\mathrm{Fr}\colon \mathbb{G}_m \longrightarrow \mathbb{G}_m$ is a $\mu_p$-torsor over $\mathbb{G}_m$ which is not étale locally trivial). To remedy this, we introduce a slight modification of the classifying stack $K(G,1)$ by changing the topology in the following way. We consider the simplicial presheaf $BG\colon X \longmapsto B(G(X))$, viewed as an object in $sPr_{\mathrm{ffqc}}(Aff)$, the model category of simplicial presheaves on the site of affine schemes endowed with the faithfully flat and quasi-compact topology ("ffqc" for short). Note that étale coverings are ffqc coverings, and therefore we have a natural full embedding

$$\mathrm{Ho}(sPr_{\mathrm{ffqc}}(\mathit{Aff})) \subset \mathrm{Ho}(sPr(\mathit{Aff})),$$

where the objects in $\mathrm{Ho}(sPr_{\mathrm{ffqc}}(\mathit{Aff}))$ are stacks satisfying the more restrictive descent condition for ffqc hypercoverings. We consider the simplicial presheaf $BG \in sPr(\mathit{Aff})$, and denote by $K_{fl}(G,1) \in \mathrm{Ho}(sPr_{\mathrm{ffqc}}(\mathit{Aff})) \subset \mathrm{Ho}(sPr(\mathit{Aff}))$ a fibrant replacement of $BG$ in the model category of stacks for the ffqc topology. It is a non-trivial statement that $K_{fl}(G,1)$ is an algebraic stack (see for instance [La-Mo, Proposition 10.13.1]). Moreover, the natural morphism $K_{fl}(G,1) \longrightarrow X$ is smooth. Indeed, we choose a smooth and surjective morphism $Y \longrightarrow K_{fl}(G,1)$, with $Y$ an affine scheme. The composition $Y \longrightarrow X$ is clearly a flat surjective morphism of finite presentation. We let $X' = Y \times^h_{K_{fl}(G,1)} X$, and consider the diagram of stacks

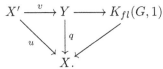

In this diagram, $v$ is a flat surjective morphism of finite presentation, because it is the base change of the trivial section $X \longrightarrow K_{fl}(G,1)$, which is flat, surjective and of finite presentation. Moreover, $u$ is a smooth morphism, because it is the base change of the smooth atlas $Y \longrightarrow K_{fl}(G,1)$. We conclude that the morphism $q$ is also smooth.

**Higher classifying stacks:** Assume now that $A$ is a sheaf of abelian groups over an affine scheme $X$ which is an algebraic space, flat and of finite presentation over $X$. We form the simplicial presheaf $B^n(A) = B(B^{n-1}(A))$, by iterating the classifying space construction. We denote by $K_{fl}(A,n) \in \mathrm{Ho}(sPr_{\mathrm{ffqc}}(\mathit{Aff})) \subset \mathrm{Ho}(sPr(\mathit{Aff}))$ a fibrant model for $B^n(A)$ with respect to the ffqc topology. It is again true that $K_{fl}(A,n)$ is an algebraic $n$-stack when $n > 1$. Indeed, $K(A,n)$ is the quotient of $X$ by the trivial action of the group stack $K(A,n-1)$. As this group stack is algebraic and smooth for $n > 1$, the quotient stack is again an algebraic stack.

**Groupoid quotients:** We describe here the standard way to construct algebraic stacks using quotients by smooth groupoid actions. We start with a simplicial object in $sPr(C)$,

$$F_* : \Delta^{\mathrm{op}} \longrightarrow sPr(\mathit{Aff}).$$

We say that $F_*$ is a *Segal groupoid* if it satisfies the following two conditions:

1. For any $n > 1$, the natural morphism

$$F_n \longrightarrow F_1 \times^h_{F_0} F_1 \times^h_{F_0} \cdots \times^h_{F_0} F_1,$$

induced by the morphism $[1] \longrightarrow [n]$ sending 0 to $i$ and 1 to $i+1$ (for $0 \leq i < n$) is an isomorphism of stacks.

2. The natural morphism
$$F_2 \longrightarrow F_1 \times^h_{F_0} F_1$$
induced by the morphism $[1] \longrightarrow [2]$ sending 0 to 0 and 1 to 1 or 2 is an isomorphism of stacks.

**Exercise 3.2.1.** Let $F_*$ be a Segal monoid object in $sPr(Aff)$, and suppose that $F_n(X)$ is a set for all $n$ and all $X$. Show that $F_*$ is the nerve of a presheaf of groupoids on $Aff$.

We now assume that $F_*$ is a Segal groupoid and moreover that all the face morphisms $F_1 \longrightarrow F_0$ are smooth morphisms between algebraic stacks. We consider the homotopy colimit of the diagram $[n] \longmapsto F_n$, and denote it by $|F_*| \in \text{Ho}(sPr(Aff))$. The stack $|F_*|$ is called the *quotient stack* of the Segal groupoid $F_*$. It can been proved that $|F_*|$ is again an algebraic stack. Moreover, if each $F_i$ is an algebraic $n$-stack, then $|F_*|$ is an algebraic $(n+1)$-stack. This is a formal way to produce higher algebraic stacks starting, say, from schemes, but this is often not the way stacks arise in practice.

An important very special case of the quotient stack construction is the case of a smooth group scheme $G$ acting on a scheme $X$. In this case we form the groupoid object $B(X, G)$ whose value in degree $n$ is $X \times G^n$, and whose transition morphisms are given by the action of $G$ on $X$. This is a groupoid object in schemes and thus can be considered as a groupoid object in sheaves, and therefore as a very special kind of Segal groupoid. The quotient stack of this Segal groupoid is denoted by $[X/G]$ and is called the quotient stack of $X$ by $G$. It is an algebraic 1-stack for which a natural smooth atlas is the natural projection $X \longrightarrow [X/G]$. It can be characterized by a universal property: morphisms of stacks $[X/G] \longrightarrow F$ are in one-to-one correspondence with morphisms of $G$-equivariant stacks $X \longrightarrow F$ (here we need to use a model category $G\text{-}sPr(Aff)$ of $G$-equivariant simplicial presheaves in order to have the correct homotopy category of $G$-equivariant stacks).

**Simplicial presentation:** Algebraic stacks can also be characterized as the simplicial presheaves represented by a certain kind of simplicial schemes. For this, we let $X_*$ be a simplicial object in the category of schemes. For any finite simplicial set $K$ (finite here means generated by a finite number of cells), we can form $X^K_*$, which is the scheme of morphisms from $K$ to $X_*$. It is, by definition, the equalizer of the two natural morphisms

$$\prod_{[n]} X^{K_n}_n \rightrightarrows \prod_{[p] \longrightarrow [q]} X^{K_q}_p.$$

This equalizer exists as a scheme when $K$ is finite (because it then only involves finite limits).

A simplicial scheme $X_*$ is then called a *weak smooth groupoid* if, for any $0 \le k \le n$, the natural morphism

$$X_n = X_*^{\Delta^n} \longrightarrow X_*^{\Lambda^{n,k}}$$

is a smooth and surjective morphism of schemes (surjective here has to be understood pointwise, but as the morphism is smooth this is equivalent to saying that it induces an epimorphism on the corresponding sheaves). A weak smooth groupoid $X_*$ is moreover *n-truncated* if, for any $k > n + 1$, the natural morphism

$$X_k = X_*^{\Delta^k} \longrightarrow X_*^{\partial \Delta^k}$$

is an isomorphism.

It is then possible to prove that a stack $F$ is an algebraic $n$-stack if there exists an $n$-truncated weak smooth groupoid $X_*$ and an isomorphism in $\mathrm{Ho}(sPr(Aff))$ $F \simeq X_*$. We refer to [Pr] for details.

**Some famous algebraic 1-stacks:** We review here two famous examples of algebraic 1-stacks, namely the stack of smooth and proper curves and the stack of vector bundles on a curve. We refer to [La-Mo] for more details.

For $X \in Aff$ an affine scheme, we let $\mathcal{M}_g(X)$ be the full subgroupoid of sheaves $F$ on $Aff/X$ such that the corresponding morphism of sheaves $F \longrightarrow X$ is representable by a smooth and proper curve of genus $g$ over $X$ (i.e., $F$ is itself a scheme, and the morphism $F \longrightarrow X$ is smooth, proper, with geometric fibers being connected curves of genus $g$). For $Y \longrightarrow X$ in $Aff$, we have a restriction functor from sheaves on $Aff/X$ to sheaves on $Aff/Y$, and this defines a natural functor of groupoids

$$\mathcal{M}_g(X) \longrightarrow \mathcal{M}_g(Y).$$

This defines a presheaf of groupoids on $Aff$, and taking the nerve of these groupoids gives a simplicial presheaf denoted by $\mathcal{M}_g$. The stack $\mathcal{M}_g$ is called the *stack of smooth curves of genus $g$*. It is such that, for $X \in Aff$, $\mathcal{M}_g(X)$ is a 1-truncated simplicial set whose $\pi_0$ is the set of isomorphism classes of smooth proper curves of genus $g$ over $X$, and whose $\pi_1$ at a given curve is its automorphism group. It is a well-known theorem that $\mathcal{M}_g$ is an algebraic 1-stack which is smooth and of finite presentation over $\mathrm{Spec}\,\mathbb{Z}$. This stack is even Deligne–Mumford, that is, the diagonal morphism $\mathcal{M}_g \longrightarrow \mathcal{M}_g \times \mathcal{M}_g$ is unramified (i.e., locally a closed immersion for the étale topology). Equivalently, this means that there exists an atlas $X \longrightarrow \mathcal{M}_g$ which is étale rather than only smooth.

Another very important and famous example of an algebraic 1-stack is the stack of $G$-bundles on some smooth projective curve $C$ (say, over some base field $k$). Let $G$ be a smooth affine algebraic group over $k$. We start by considering the stack $BG$, which is a stack over $\mathrm{Spec}\,k$. It is the quotient stack $[\mathrm{Spec}\,k/G]$ for the trivial action of $G$ on $\mathrm{Spec}\,k$. As $G$ is a smooth algebraic group, this stack is an algebraic

1-stack. When $C$ is a smooth and proper curve over $\operatorname{Spec} k$, we can consider the stack of morphisms (of stacks over $\operatorname{Spec} k$)

$$\operatorname{Bun}_G(C) = \mathbb{R}\underline{\mathcal{H}om}_{A\!f\!f/\operatorname{Spec} k}(C, BG),$$

which by definition is the stack of principal $G$-bundles on $C$. By definition, for $X \in A\!f\!f$, $\operatorname{Bun}_G(C)(X)$ is a 1-truncated simplicial set whose $\pi_0$ is the set of isomorphism classes of principal $G$-bundles on $C$ and whose $\pi_1$ at a given bundle is its automorphism group. It is also a well-known theorem that the stack $\operatorname{Bun}_G(C)$ is an algebraic 1-stack, which is smooth and locally of finite presentation over $\operatorname{Spec} k$. However, this stack is not quasi-compact and is only a countable union of quasi-compact open substacks.

**Higher linear stacks:** Let $X = \operatorname{Spec} A$ be an affine scheme and $E$ be a positively graded cochain complex of $A$-modules. We assume that $E$ is perfect, i.e., it is quasi-isomorphic to a bounded complex of projective $A$-modules of finite type. We define a stack $\mathbb{V}(E)$ over $X$ in the following way. For every commutative $A$-algebra $B$, we set

$$\mathbb{V}(E)(B) = \operatorname{Map}(E, B),$$

where Map denotes the mapping space of the model category of complexes of $A$-modules. More explicitly, $\mathbb{V}(E)(B)$ is the simplicial set whose set of $n$-simplicies is the set $\operatorname{Hom}(Q(E) \otimes_A C_*(\Delta^n, A), B)$. Here $Q(E)$ is a cofibrant resolution of $E$ in the model category of complexes of $A$-modules (for the projective model structure, for which equivalences are quasi-isomorphisms and fibrations are epimorphisms), $C_*(\Delta^n, A)$ is the homology complex of the simplicial set $\Delta^n$ with coefficients in $A$, and the Hom is taken in the category of complexes of $A$-modules. In other words, $\mathbb{V}(E)(B)$ is the simplicial set obtained from the complex $\underline{\operatorname{Hom}}^*(Q(E), B)$ by the Dold–Kan correspondence. When $B$ varies in the category of commutative $A$-algebras, this defines a simplicial presheaf $\mathbb{V}(E)$ together with a morphism $\mathbb{V}(E) \longrightarrow X = \operatorname{Spec} A$. For every commutative $A$-algebra $B$, we have

$$\pi_i(\mathbb{V}(E)(B)) \simeq \operatorname{Ext}^{-i}(E, B).$$

It can be shown that the stack $\mathbb{V}(E)$ is an algebraic $n$-stack strongly of finite presentation over $X$, where $n$ is such that $H^i(E) = 0$ for all $i > n$, and that $\mathbb{V}(E)$ is smooth if and only if the Tor amplitude of $E$ is non-negative (i.e., $E$ is quasi-isomorphic to a complex of projective $A$-modules of finite type which is moreover concentrated in non-negative degrees). For this, we can first assume that $E$ is a bounded complex of projective modules of finite type. We then set $K = E^{\leq 0}$, the part of $E$ which is concentrated in non-positive degrees, and we have a natural morphism of complexes $E \longrightarrow K$. This morphism induces a morphism of stacks

$$\mathbb{V}(K) \longrightarrow \mathbb{V}(E).$$

By definition, $\mathbb{V}(K)$ is naturally equivalent to the affine scheme $\operatorname{Spec} A[H^0(K)]$, where $A[H^0(K)]$ denotes the free commutative $A$-algebra generated by the $A$-module $H^0(K)$. It is well known that $\mathbb{V}(H^0(k))$ is smooth over $\operatorname{Spec} A$ if and only if $H^0(K)$ is projective and of finite type. This is equivalent to saying that $E$ has non-negative Tor amplitude. The only thing to check is then that the natural morphism

$$\mathbb{V}(K) \longrightarrow \mathbb{V}(E)$$

is $(n-1)$-algebraic and smooth. But this follows by induction on $n$, as this morphism is locally on $\mathbb{V}(E)$ of the form $Y \times \mathbb{V}(L) \longrightarrow Y$, for $L$ the homotopy cofiber (i.e., the cone) of the morphism $E \longrightarrow K$. This homotopy cofiber is itself quasi-isomorphic to $E^{>0}[1]$, and thus is a perfect complex of non-negative Tor amplitude with $H^i(L) = 0$ for $i > n-1$.

**Exercise 3.2.2.** Let $X = \mathbb{A}^1 = \operatorname{Spec} \mathbb{Z}[T]$ and let $E$ be the complex of $\mathbb{Z}[T]$-modules given by

$$0 \longrightarrow \mathbb{Z}[T] \xrightarrow{\times T} \mathbb{Z}[T] \longrightarrow 0,$$

concentrated in degrees 1 and 2. Show that $\mathbb{V}(E)$ is an algebraic 2-stack such that the sheaf $\pi_1(\mathbb{V}(E))$ is not representable by any affine scheme (it is in fact not representable by any algebraic space).

**The algebraic 2-stack of abelian categories:** This is a non-trivial example of an algebraic 2-stack. The material is taken from [An]. For a commutative ring $A$, we consider the following category $Ab(A)$. Its objects are abelian $A$-linear categories which are equivalent to the category $R\text{-}Mod$ of left $R$-modules for some associative $A$-algebra $R$ which is projective and of finite type as an $A$-module. The morphisms in $Ab(A)$ are the $A$-linear equivalences of categories. For a morphism of commutative rings $A \longrightarrow B$, we have a functor

$$Ab(A) \longrightarrow Ab(B)$$

sending an abelian category $\mathcal{C}$ to $\mathcal{C}^{B/A}$, the category of $B$-modules in $\mathcal{C}$. Precisely $\mathcal{C}^{B/A}$ can be taken to be the category of all $A$-linear functors from $BB$, the $A$-linear category with a unique object and $B$ as its $A$-algebra of endomorphisms, to $\mathcal{C}$. This defines a presheaf of categories $A \longmapsto Ab(A)$ on *Aff*. Taking the nerves of these categories, we obtain a simplicial presheaf $\mathbf{Ab} \in sPr(Aff)$. The simplicial presheaf $\mathbf{Ab}$ is not a stack, but we still consider it as an object in $\operatorname{Ho}(sPr(Aff))$. The main result of [An] states that $\mathbf{Ab}$ is an algebraic 2-stack which is locally of finite presentation.

**The algebraic $n$-stack of $[n, 0]$-perfect complexes:** For a commutative ring $A$, we consider a category $P(A)$ defined as follows. Its objects are the cofibrant complexes of $A$-modules (for the projective model structure) which are perfect (i.e., quasi-isomorphic to a bounded complex of projective modules of finite type). The morphisms in $P(A)$ are the quasi-isomorphisms of complexes of $A$-modules. For a morphism of commutative rings $A \longrightarrow B$, we have a base-change functor

$$- \otimes_A B \colon P(A) \longrightarrow P(B).$$

This does not however define, stricly speaking, a presheaf of categories, as the base-change functors are only compatible with composition up to a natural isomorphism. In other words, $A \longmapsto P(A)$ is only a weak functor from *Comm* to the 2-category of categories. Fortunately, there exists a standard procedure to replace any weak functor by an equivalent strict functor: it consists in replacing $P$ by the presheaf of cartesian sections of the Grothendieck construction $\int P \longrightarrow Comm$ (see [SGA1]). Thus, we define a new category $P'(A)$ whose objects consist of the following data:

1. For any commutative $A$-algebra $B$, an object $E_B \in P(B)$.

2. For any commutative $A$-algebra $B$ and any commutative $B$-algebra $C$, an isomorphism in $P'(C)$,

$$\phi_{B,C} \colon E_B \otimes_B C \simeq E_C.$$

We require moreover that, for any commutative $A$-algebra $B$, any commutative $B$-algebra $C$, and any commutative $C$-algebra $D$, the two possible isomorphisms

$$\phi_{C,D} \circ (\phi_{B,C} \otimes_C D) \colon (E_B \otimes_B C) \otimes_C D \simeq E_B \otimes_B D \longrightarrow E_D$$

$$\phi_{B,D} \colon E_B \otimes_B D \longrightarrow E_D$$

are equal. The morphisms in $P'(A)$ are simply taken to be families of morphisms $E_B \longrightarrow E'_B$ which commute with the collections $\phi_{B,C}$ and $\phi'_{B,C}$.

With these definitions, $A \longmapsto P'(A)$ is a functor *Comm* $\longrightarrow$ *Cat*, and there is moreover an equivalence of lax functors $P' \longrightarrow P$. We compose the functor $P'$ with the nerve construction and get a simplicial presheaf **Perf** on *Aff*. It can be proved that the simplicial presheaf **Perf** is a stack in the sense of Definition 2.1.3 (1). This is not an obvious result (see for instance [H-S] for a proof), and can be reduced to the well-known flat cohomological descent for quasi-coherent complexes. It can also be proved that, for $X = \operatorname{Spec} A \in Aff$, the simplicial set **Perf**$(X)$ satisfies the following properties:

1. The set $\pi_0(\mathbf{Perf}(X))$ is in a natural bijection with the set of quasi-isomorphism classes of perfect complexes of $A$-modules.

2. For $x \in \mathbf{Perf}(X)$ corresponding to a perfect complex $E$, we have

$$\pi_1(\mathbf{Perf}(X), x) \simeq \operatorname{Aut}(E),$$

where the automorphism group is taken in the derived category $D(A)$ of the ring $A$.

3. For $x \in \mathbf{Perf}(X)$ corresponding to a perfect complex $E$, we have

$$\pi_i(\mathbf{Perf}(X), x) \simeq \mathrm{Ext}^{1-i}(E, E)$$

for any $i > 1$. Again, these Ext groups are computed in the triangulated category $D(A)$.

For any $n \geq 0$ and $a \leq b$ with $b - a = n$, we can define a subsimplicial presheaf $\mathbf{Perf}^{[a,b]} \subset \mathbf{Perf}$ which consists of all perfect complexes of Tor amplitude contained in the interval $[a, b]$ (i.e., complexes quasi-isomorphic to a complex of projective modules of finite type concentrated in degrees $[a, b]$). It can be proved that the substacks $\mathbf{Perf}^{[a,b]}$ form an open covering of $\mathbf{Perf}$. Moreover, $\mathbf{Perf}^{[a,b]}$ is an algebraic $(n + 1)$-stack which is locally of finite presentation. This way, even though $\mathbf{Perf}$ is not, strictly speaking, an algebraic stack (because it is not an $n$-stack for any $n$), it is an increasing union of open algebraic substacks. We say that $\mathbf{Perf}$ is *locally algebraic*. The fact that $\mathbf{Perf}^{[a,b]}$ is an algebraic $(n + 1)$-stack is not easy either. We refer to [To-Va] for a complete proof.

**Exercise 3.2.3.**   1. Show how to define a stack $\mathbf{MPerf}$ of morphisms between perfect complexes, whose value at $X \in \mathit{Aff}$ is equivalent to the nerve of the category of quasi-isomorphisms in the category of morphisms between perfect complexes over $X$.

2. Show that the morphism *source* and *target* define an algebraic morphism of stacks
$$\pi\colon \mathbf{MPerf} \longrightarrow \mathbf{Perf} \times \mathbf{Perf}.$$

   (Here you will need the following result of homotopical algebra: If $M$ is a model category and $\mathrm{Mor}(M)$ denotes the model category of morphisms, then the homotopy fiber of the source and target map $N(w\mathrm{Mor}(M)) \longrightarrow N(wM) \times N(wM)$, taken at a point $(x, y)$, is naturally equivalent to the mapping space $\mathrm{Map}(x, y)$.)

3. Show that the morphism $\pi$ is locally smooth near any point corresponding to a morphism $E \longrightarrow E'$ of perfect complexes such that $\mathrm{Ext}^i(E, E') = 0$ for all $i > 0$.

## 3.3   Coarse moduli spaces and homotopy sheaves

The purpose of this part is to show that algebraic $n$-stacks strongly of finite presentation can be approximated by schemes by means of some *dévissage*. The existence of this approximation has several important consequences about the behaviour of algebraic $n$-stacks, such as the existence of virtual coarse moduli spaces or homotopy group schemes. Conceptually, the results of this part show that algebraic $n$-stacks are not that far from being schemes or algebraic spaces, and that for many purposes they behave like *convergent series of schemes*.

**Convention:** Throughout this part, all algebraic $n$-stacks will be strongly of finite presentation over some affine base scheme $\operatorname{Spec} k$ (for $k$ some commutative ring).

The key notion is that of *total gerbe*, whose precise definition is as follows.

**Definition 3.3.1.** Let $F$ be an algebraic $n$-stack. We say that $F$ is a *total (n-)gerbe* if for all $i > 0$ the natural projection

$$I_F^{(i)} = \mathbb{R}\underline{\operatorname{Hom}}(S^i, F) \longrightarrow F$$

is a flat morphism.

In the previous definition, $I_F^{(i)}$ is called the *$i$-th inertia stack* of $F$. Note that, when $F$ is a 1-stack, $I_F^{(1)}$ is equivalent to the inertia stack of $F$ in the usual sense. In particular, for an algebraic 1-stack, being a total gerbe in the sense of Definition 3.3.1 is equivalent to the fact that the projection morphism

$$F \times_{F \times F}^h F \longrightarrow F$$

is flat, and thus equivalent to the usual notion of gerbes for algebraic 1-stacks (see [La-Mo, Definition 3.15]).

**Proposition 3.3.2.** *Let $F$ be an algebraic $n$-stack which is a total gerbe. Then the following conditions are satisfied:*

1. *If $M(F)$ is the sheaf associated to $\pi_0(F)$ for the flat (ffqc) topology, then $M(F)$ is represented by an algebraic space and the morphism $F \longrightarrow M(F)$ is flat and of finite presentation.*

2. *For any $X \in Aff$ and any morphism $x \colon X \longrightarrow F$, $\pi_i^{fl}(F, x)$, the sheaf on $X$ associated to $\pi_i(F, x)$ with respect to the ffqc topology, is an algebraic space, flat, and of finite presentation over $X$.*

*Proof:* Condition 1 follows from a well-known theorem of Artin, ensuring representability by algebraic spaces of quotients of schemes by flat equivalence relations. The argument goes as follows. We choose a smooth atlas $X \longrightarrow F$ with $X$ an affine scheme, and we let $X_1 = X \times_F^h X$. We define $R \subset X \times X$, the sub-ffqc-sheaf image of $X_1 \longrightarrow X \times X$, which defines an equivalence relation on $X$. Clearly, $M(F)$ is isomorphic to the quotient ffqc-sheaf $(X/R)^{fl}$. We now prove the following two properties:

1. The sheaf $R$ is an algebraic space.

2. The two projections $R \longrightarrow X$ are smooth.

In order to prove property 1, we consider the natural projection $X_1 \longrightarrow R$. Let $x, y \colon Y \longrightarrow R \subset X \times X$ be morphisms with $Y$ affine. Then $X_1 \times_R^h Y$ is equivalent to $\Omega_{x,y} F \simeq Y \times_F^h Y$, the stack of paths from $x$ to $y$. As the objects $x$ and $y$ are locally (for the flat topology) equivalent on $Y$ because $(x, y) \in R$, the stack $\Omega_{x,y}F \simeq Y \times_F^h Y$ is algebraic and locally (for the flat topology on $Y$) equivalent to the loop stack $\Omega_x F$, defined by the homotopy cartesian square

$$
\begin{array}{ccc}
\Omega_x F & \longrightarrow & I_F^{(1)} \\
\downarrow & & \downarrow \\
Y & \longrightarrow & F.
\end{array}
$$

By hypothesis on $F$, we deduce that $X_1 \times_R^h Y \longrightarrow Y$ is flat, surjective and of finite presentation. As this is true for any $Y \longrightarrow R$, we have that the morphism of stacks $X_1 \longrightarrow R$ is surjective, flat and finitely presented. If $U \longrightarrow X_1$ is a smooth atlas, we have that the sheaf $R$ is isomorphic to the quotient ffqc-sheaf

$$
R \simeq \mathrm{Colim}\,(U \times_{X \times X} U \rightrightarrows U),
$$

and, by what we have just seen, the projections $U \times_{X \times X} U \longrightarrow U$ are flat and finitely presented morphisms of affine schemes. By [La-Mo, Corollary 10.4], we have that $R$ is an algebraic space.

We now consider property 2. For this, we consider the diagram

$$
U \longrightarrow R \longrightarrow X.
$$

The first morphism is a flat and finitely presented cover, and the composition of the two morphisms equals the composition $U \longrightarrow X_1 \longrightarrow X$, and is thus smooth. Hence $R \longrightarrow X$ is locally (for the flat finitely presented topology on $R$) a smooth morphism, and therefore it is smooth. This finishes the proof of the first part of the proposition, as $X \longrightarrow M(F)$ is now a smooth atlas, showing that $M(F)$ is an algebraic space.

To prove the second statement of the proposition, we will use (1) applied to certain stacks of iterated loops. We let $x \colon X \longrightarrow F$ and consider the loop stack $\Omega_x F$ of $F$ at $x$, defined by

$$
\Omega_x F = X \times_F^h X.
$$

In the same way, we have the iterated loop stacks

$$
\Omega_x^{(i)} F = \Omega_x(\Omega_x^{(i-1)} F).
$$

Note that we have homotopy cartesian squares

$$
\begin{array}{ccc}
\Omega_x^{(i)} F & \longrightarrow & I_F^{(i)} \\
\downarrow & & \downarrow \\
X & \longrightarrow & F,
\end{array}
$$

showing that $\Omega_x^{(i)} F \longrightarrow X$ is flat for any $i$. Moreover, for any $Y \in Aff$ and any $s \colon Y \longrightarrow \Omega_x^{(i)} F$, we have a homotopy cartesian square

$$
\begin{array}{ccc}
\Omega_s^{(j)} \Omega_x^{(i)} F & \longrightarrow & I_{\Omega_x^{(i)} F}^{(j)} \\
\downarrow & & \downarrow \\
Y & \longrightarrow & \Omega_x^{(i)} F.
\end{array}
$$

Now, as $\Omega_x^{(i)} F$ is a group object over $X$, we have isomorphisms of stacks over $Y$,

$$\Omega_s^{(j)} \Omega_x^{(i)} F \simeq \Omega_x^{(j+i)} F \times_X Y$$

obtained by translating along the section $s$. Therefore, we have that

$$I_{\Omega_x^{(i)} F}^{(j)} \longrightarrow \Omega_x^{(i)} F$$

is flat for any $i$ and any $j$. We can therefore apply (1) to the stacks $\Omega_x^{(i)} F$. As we have

$$M(\Omega_x^{(i)} F) \simeq \pi_i^{fl}(F, x),$$

this gives that the sheaves $\pi_i^{fl}(F, x)$ are algebraic spaces. Moreover, the morphism $\Omega_x^{(i)} F \longrightarrow \pi_i^{fl}(F, x)$ is flat, surjective and of finite presentation, showing that so is $\pi_i^{fl}(F, x)$ as an algebraic space over $X$. □

**Exercise 3.3.3.** 1. Let $f \colon F \longrightarrow F'$ be a morphism of finite presentation between algebraic stacks strongly of finite presentation over some affine scheme. Assume that $F'$ is reduced. Show that there exists a non-empty open substack $U \subset F'$ such that the base-change morphism $F \times_{F'}^h U \longrightarrow U$ is flat (use smooth atlases and the generic flatness theorem statement that the result is true when $F$ and $F'$ are affine schemes).

2. Deduce from (1) that, if $F$ is a reduced algebraic stack strongly of finite presentation over some affine scheme, then $F$ has a non-empty open substack $U \subset F$ which is a total gerbe in the sense of Definition 3.3.1.

The previous exercise, together with Proposition 3.3.2, has the following important consequence:

**Corollary 3.3.4.** *Let $F$ be an algebraic stack strongly of finite presentation over some affine scheme $X$. There exists a finite sequence of closed substacks*

$$\emptyset \subset F_r \subset F_{r-1} \subset \cdots \subset F_1 \subset F_0 = F$$

*such that each $F_i - F_{i+1}$ is a total gerbe. We can moreover choose the $F_i$ with the following properties:*

1. *For all $i$, the ffqc-sheaf $M(F_i - F_{i+1})$ is a scheme of finite type over $X$.*

2. *For all $i$, all affine schemes $Y$, all morphisms $y\colon Y \longrightarrow (F_i - F_{i+1})$, and all $j > 0$, the ffqc-sheaf $\pi_j(F_i - F_{i+1}, y)$ is a flat algebraic space of finite presentation over $Y$.*

In other words, any algebraic stack $F$ strongly finitely presented over some affine scheme gives rise to several schemes $M(F_i - F_{i+1})$, which are stratified pieces of the non-existing coarse moduli space for $F$. Over each of these schemes, locally for the étale topology, we have the flat groups $\pi_j(F_i - F_{i+1})$. Therefore, up to a stratification, the stack $F$ behaves very much like a homotopy type whose homotopy groups would be represented by schemes (or algebraic spaces).

**Exercise 3.3.5.** Recall that an algebraic stack is *Deligne–Mumford* if it possesses an étale atlas (rather than simply smooth).

1. Let $F$ be an algebraic stack which is étale over an affine scheme $X$. Prove that $F$ is Deligne–Mumford and that $F$ is a total gerbe. Show also that the projection $F \longrightarrow M(F)$ is an étale morphism.

2. Let $F$ be a Deligne–Mumford stack and $p\colon F \longrightarrow t_{\leq 1}(F)$ be its 1-truncation. Show that $t_{\leq 1}(F)$ is itself a Deligne–Mumford 1-stack and that $p$ is étale.

# Lecture 4

# Simplicial commutative algebras

In this lecture we review the homotopy theory of simplicial commutative rings. It will be used throughout the next lectures in order to define and study the notion of derived schemes and derived stacks.

## 4.1 Quick review of the model category of commutative simplicial algebras and modules

We let $sComm$ be the category of simplicial objects in $Comm$, that is, of simplicial commutative algebras. For $A \in sComm$ a simplicial commutative algebra, we let $sA\text{-}Mod$ be the category of simplicial $A$-modules. Recall that an object in $sA\text{-}Mod$ is the data of a simplicial abelian group $M_n$ together with $A_n$-module structures on $M_n$ in such a way that the transition morphisms $M_n \longrightarrow M_m$ are morphisms of $A_n$-modules (for the $A_n$-module structure on $M_m$ induced by the morphism $A_n \longrightarrow A_m$). We will say that a morphism in $sComm$ or in $sA\text{-}Mod$ is an equivalence (resp. a fibration), if the morphism induced on the underlying simplicial sets is so. It is well known that this defines model category structures on $sComm$ and $sA\text{-}Mod$. These model categories are cofibrantly generated, proper and cellular.

For any simplicial commutative algebra $A$, let $\pi_*(A) = \oplus_n \pi_n(A)$. We do not specify base points since the underlying simplicial set of a simplicial algebra is a simplicial abelian group, and thus its homotopy groups do not depend on the base point (by convention we will take $0$ as base point). The graded abelian group $\pi_*(A)$ has a natural structure of a graded commutative (in the graded sense) algebra. The multiplication of two elements $a \in \pi_n(A)$ and $b \in \pi_m(A)$ is defined as follows. We represent $a$ and $b$ by morphisms of pointed simplicial sets

$$a\colon S^n = (S^1)^{\wedge n} \longrightarrow A, \qquad b\colon S^m = (S^1)^{\wedge n} \longrightarrow A,$$

I. Moerdijk and B. Toën, *Simplicial Methods for Operads and Algebraic Geometry*, Advanced Courses in Mathematics - CRM Barcelona, DOI 10.1007/978-3-0348-0052-5_12, © Springer Basel AG 2010

where $S^1$ is a model for the pointed simplicial circle. We then consider the induced morphism

$$a \otimes b \colon S^n \times S^m \longrightarrow A \times A \longrightarrow A \otimes A.$$

Composing with the multiplication in $A$ we get a morphism of simplicial sets

$$ab \colon S^n \times S^m \longrightarrow A.$$

This last morphism sends $S^n \times *$ and $* \times S^m$ to the base point $0 \in A$. Therefore, it factorizes as a morphism

$$S^n \wedge S^m \simeq S^{n+m} \longrightarrow A.$$

As the left-hand side has the homotopy type of $S^{n+m}$, we obtain a morphism

$$S^{n+m} \longrightarrow A$$

which gives an element $ab \in \pi_{n+m}(A)$. This multiplication is associative, unital and graded commutative. In the same way, if $A$ is a simplicial commutative algebra and $M$ a simplicial $A$-module, $\pi_*(M) = \oplus_n \pi_n(M)$ has a natural structure of a graded $\pi_*(A)$-module.

For a morphism of simplicial commutative rings $f \colon A \longrightarrow B$, we have an adjunction

$$- \otimes_A B \colon sA\text{-}Mod \longrightarrow sB\text{-}Mod, \qquad sA\text{-}Mod \longleftarrow sB\text{-}Mod \colon f^*,$$

where the right adjoint $f^*$ is the forgetful functor. This adjunction is a Quillen adjunction which is moreover a Quillen equivalence if $f$ is an equivalence of simplicial algebras. The left derived functor of $- \otimes_A B$ will be denoted by

$$- \otimes_A^{\mathbb{L}} B \colon \mathrm{Ho}(sA\text{-}Mod) \longrightarrow \mathrm{Ho}(sB\text{-}Mod).$$

Finally, a (non-simplicial) commutative ring will always be considered as a constant simplicial commutative ring and thus as an object in $sComm$. This induces a fully faithful functor $Comm \longrightarrow sComm$ which induces a fully faithful embedding

$$Comm \longrightarrow \mathrm{Ho}(sComm)$$

on the level of the homotopy category. This last functor possesses a left adjoint

$$\pi_0 \colon \mathrm{Ho}(sComm) \longrightarrow Comm.$$

In the same manner, if $A \in sComm$, then any (non-simplicial) $\pi_0(A)$-module can be considered as a constant simplicial $A$-module, and thus as an object in $sA$-Mod. This also defines a full embedding

$$\pi_0(A)\text{-}Mod \longrightarrow \mathrm{Ho}(sA\text{-}Mod)$$

which admits a left adjoint

$$\pi_0 \colon \mathrm{Ho}(sA\text{-}Mod) \longrightarrow \pi_0(A)\text{-}Mod.$$

## 4.2 Cotangent complexes

We start by recalling the notion of trivial square-zero extension of a commutative ring by a module. For any commutative ring $A$ and any $A$-module $M$, we define another commutative ring $A \oplus M$. The underlying abelian group of $A \oplus M$ is the direct sum of $A$ and $M$, and the multiplication is defined by the following formula:

$$(a, m) \cdot (a', m') = (aa', am' + a'm).$$

The commutative ring $A \oplus M$ is called the *trivial square-zero extension* of $A$ by $M$. It is an augmented $A$-algebra by the natural morphisms

$$A \longrightarrow A \oplus M \longrightarrow A,$$

sending respectively $a$ to $(a, 0)$ and $(a, m)$ to $a$. The main property of the ring $A \oplus M$ is that the set of sections of the projection $A \oplus M \longrightarrow A$ (as morphisms or rings) is in natural bijection with the set $\mathrm{Der}(A, M)$ of derivations from $A$ to $M$ (this can be taken as a definition of $\mathrm{Der}(A, M)$). A standard result states that the functor

$$A\text{-}Mod \longrightarrow Set$$

sending $M$ to $\mathrm{Der}(A, M)$ is corepresented by an $A$-module $\Omega_A^1$, the $A$-module of Kähler differentials on $A$.

**Exercise 4.2.1.** Let $A$ be a commutative ring and consider the category $Comm/A$ of commutative rings augmented over $A$. Show that $M \longmapsto A \oplus M$ defines an equivalence of categories between the category $A$-$Mod$ of $A$-modules and the category $Ab(Comm/A)$ of abelian group objects in $Comm/A$. Show that, under this equivalence, $\Omega_A^1$ is the free abelian group object in $Comm/A$ over a single generator.

The generalization of the above notion to the context of simplicial commutative rings leads to the notion of cotangent complexes and André–Quillen homology (and cohomology). We let $A$ be a simplicial commutative ring and $M \in sA$-$Mod$ be a simplicial $A$-module. By applying the construction of the trivial square-zero extension levelwise for each $A_n$ and each $M_n$, we obtain a new simplicial commutative ring $A \oplus M$ together with two morphisms

$$A \longrightarrow A \oplus M \longrightarrow A.$$

The model category $sComm$ is a simplicial model category. We will denote by $\underline{\mathrm{Hom}}$ its simplicial Hom sets and by $\mathbb{R}\underline{\mathrm{Hom}}$ its derived version (i.e., $\mathbb{R}\underline{\mathrm{Hom}}(A, B) = \underline{\mathrm{Hom}}(Q(A), B)$, where $Q(A)$ is a cofibrant model for $A$). The simplicial set $\mathbb{R}\mathrm{Der}(A, M)$ of derived derivations from $A$ to $M$ is by definition the homotopy fiber of the natural morphism

$$\mathbb{R}\underline{\mathrm{Hom}}(A, A \oplus M) \longrightarrow \mathbb{R}\underline{\mathrm{Hom}}(A, A)$$

taken at the identity. On the other hand, we have

$$\mathbb{R}\mathrm{Der}(A, M) \simeq \underline{\mathbb{R}\mathrm{Hom}}_{sComm/A}(A, A \oplus M),$$

where $\underline{\mathrm{Hom}}_{sComm/A}$ denotes the simplicial Hom sets of the model category of commutative simplicial rings over $A$. It is a well-known result that the functor

$$\mathbb{R}\mathrm{Der}(A, -) \colon \mathrm{Ho}(sA\text{-}Mod) \longrightarrow \mathrm{Ho}(sSet)$$

is corepresented by a simplicial $A$-module $\mathbb{L}_A$ called the *cotangent complex* of $A$ (see for example [Q, Go-Ho]). One possible construction of $\mathbb{L}_A$ is as follows. We start by considering a cofibrant replacement $Q(A) \longrightarrow A$ for $A$. We then apply the construction of Kähler differentials levelwise for $Q(A)$ to get a simplicial $Q(A)$-module $\Omega^1_{Q(A)}$. We then set

$$\mathbb{L}_A = \Omega^1_{Q(A)} \otimes^{\mathbb{L}}_{Q(A)} A \in \mathrm{Ho}(sA\text{-}Mod).$$

In this way, $A \longmapsto \mathbb{L}_A$ is the left derived functor of $A \longmapsto \Omega^1_A$. We note that, by adjunction, we always have

$$\pi_0(\mathbb{L}_A) \simeq \Omega^1_{\pi_0(A)}.$$

The cotangent complex is functorial in $A$. Therefore, for any morphism of simplicial commutative rings $A \longrightarrow B$ we have a morphism $\mathbb{L}_A \longrightarrow \mathbb{L}_B$ in $\mathrm{Ho}(sA\text{-}Mod)$. By adjunction, this morphism can also be considered as a morphism in $\mathrm{Ho}(sB\text{-}Mod)$,

$$\mathbb{L}_A \otimes^{\mathbb{L}}_A B \longrightarrow \mathbb{L}_B.$$

The homotopy cofiber of this morphism will be denoted by $\mathbb{L}_{B/A}$ and called the *relative cotangent complex* of $B$ over $A$.

An important fact concerning cotangent complexes is that they can be used in order to describe Postnikov invariants of commutative rings as follows. A simplicial commutative ring $A$ is said to be *$n$-truncated* if $\pi_i(A) = 0$ for all $i > n$. The inclusion functor of the full subcategory $\mathrm{Ho}(sComm_{\leq n})$ of $n$-truncated simplicial commutative rings has a left adjoint

$$\tau_{\leq n} \colon \mathrm{Ho}(sComm) \longrightarrow \mathrm{Ho}(sComm_{\leq n}),$$

called the *$n$-th truncation functor*. These functors can easily be obtained by applying the general machinery of left Bousfield localizations to $sComm$. They are the localization functors associated to the left Bousfield localizations of $sComm$ with respect to the morphism $S^{n+1} \otimes \mathbb{Z}[T] \longrightarrow \mathbb{Z}[T]$. For any $A \in \mathrm{Ho}(sComm)$, we then have a Postnikov tower

$$A \longrightarrow \ldots \longrightarrow \tau_{\leq n}(A) \longrightarrow \tau_{\leq n-1}(A) \longrightarrow \ldots \longrightarrow \tau_{\leq 0}(A) = \pi_0(A).$$

It can be proved that for any $n > 0$ there is a homotopy pullback square

$$\begin{array}{ccc}
\tau_{\leq n}(A) & \longrightarrow & \tau_{\leq n-1}(A) \\
\downarrow & & \downarrow {\scriptstyle 0} \\
\tau_{\leq n-1}(A) & \xrightarrow{\;k_n\;} & \tau_{\leq n-1}(A) \oplus \pi_n(A)[n+1],
\end{array}$$

where $\pi_n(A)[n+1]$ is the simplicial $A$-module $S^{n+1} \otimes \pi_n(A)$ (i.e., the $(n+1)$-suspension of $\pi_n(A)$), 0 stands for the trivial derivation, and $k_n$ is a certain derived derivation with values in $\pi_n(A)[n+1]$. This derivation is an element in $[\mathbb{L}_{\tau_{\leq n-1}(A)}, \pi_n(A)[n+1]]$ which is by definition the $n$-th Postnikov invariant of $A$. This element completely determines the simplicial commutative ring $\tau_{\leq n}(A)$ from $\tau_{\leq n-1}(A)$ and the $\pi_0(A)$-module $\pi_n(A)$. It is non-zero precisely when the projection $\tau_{\leq n}(A) \longrightarrow \tau_{\leq n-1}(A)$ has no sections (in Ho($sComm$)). It is zero precisely when $\tau_{\leq n}(A)$ is equivalent (as an object over $\tau_{\leq n-1}(A)$) to $\tau_{\leq n-1}(A) \oplus \pi_n(A)[n]$.

**Exercise 4.2.2.** 1. Let $A$ be a simplicial commutative ring with $\pi_0(A)$ being isomorphic either to $\mathbb{Z}$, $\mathbb{Q}$ or $\mathbb{Z}/p$. Show that the natural projection $A \longrightarrow \pi_0(A)$ has a section in Ho($sComm$). Show moreover that this section is unique when $\pi_0(A)$ is either $\mathbb{Z}$ or $\mathbb{Q}$.

2. Give an example of a simplicial commutative ring $A$ such that the natural projection $A \longrightarrow \pi_0(A)$ has no sections in Ho($sComm$).

## 4.3 Flat, smooth and étale morphisms

We arrive at the three fundamental notions of flat, smooth and étale morphisms of commutative simplicial rings. The material of this section is less standard than the one of the previous section and thus we refer to [HAGII, §2.2.2] for details.

**Definition 4.3.1.** Let $f\colon A \longrightarrow B$ be a morphism of simplicial commutative rings.

1. The morphism $f$ is *homotopically of finite presentation* if, for any filtered system of commutative simplicial $A$-algebras $\{C_\alpha\}$, the natural map

$$\mathrm{Colim}_\alpha \, \mathbb{R}\underline{\mathrm{Hom}}_{A/sComm}(C_\alpha, B) \longrightarrow \mathbb{R}\underline{\mathrm{Hom}}_{A/sComm}(\mathrm{Colim}_\alpha \, C_\alpha, B)$$

is an equivalence.

2. The morphism $f$ is *flat* if the base-change functor

$$- \otimes_A^{\mathbb{L}} B \colon \mathrm{Ho}(sA\text{-}Mod) \longrightarrow \mathrm{Ho}(sB\text{-}Mod)$$

commutes with homotopy pullbacks.

3. The morphism $f$ is *formally étale* if $\mathbb{L}_{B/A} \simeq 0$.

4. The morphism $f$ is *formally smooth* if for any simplicial $B$-module $M$ with $\pi_0(M) = 0$ we have $[\mathbb{L}_{B/A}, M] = 0$.

5. The morphism $f$ is *smooth* if it is formally smooth and homotopically of finite presentation.

6. The morphism $f$ is *étale* if it is formally étale and homotopically of finite presentation.

7. The morphism $f$ is a *Zariski open immersion* if it is flat, homotopically of finite presentation and if moreover the natural morphism $B \otimes^{\mathbb{L}}_A B \longrightarrow B$ is an equivalence.

All these notions of morphisms are stable under composition in $\mathrm{Ho}(sComm)$. They are also stable under homotopy cobase change in the sense that, if a morphism $f \colon A \longrightarrow B$ is homotopically of finite presentation (resp. flat, formally étale, etc.), then for any $A \longrightarrow A'$ the induced morphism $A' \longrightarrow A' \otimes^{\mathbb{L}}_A B$ is again homotopically of finite presentation (resp. flat, formally étale, etc.).

Here follows a sample of standard results concerning the above notions.

- A Zariski open immersion is étale, an étale morphism is smooth, and a smooth morphism is flat.

- A morphism $A \longrightarrow B$ is flat (resp. étale, smooth, a Zariski open immersion) if and only if it satisfies the two following properties.

  1. The induced morphism of rings $\pi_0(A) \longrightarrow \pi_0(B)$ is flat (resp. étale, smooth, a Zariski open immersion) in the usual sense.

  2. For all $i > 0$ the induced morphism

  $$\pi_i(A) \otimes_{\pi_0(A)} \pi_0(B) \longrightarrow \pi_i(B)$$

  is bijective.

- An important direct consequence of the last point is that a morphism of (non-simplicial) commutative rings is flat (resp. étale, smooth, a Zariski open immersion) in the usual sense if and only if it is so in the sense of Definition 4.3.1.

- A morphism $A \longrightarrow B$ is homotopically of finite presentation if and only if $B$ is equivalent to a retract of a finite cellular $A$-algebra. Recall here that a finite cellular $A$-algebra is a commutative $A$-algebra $B'$ such that there exists a finite sequence

$$A = B'_0 \longrightarrow B'_1 \longrightarrow \ldots \longrightarrow B'_n = B'$$

such that for any $i$ there exists a cocartesian square of commutative simplicial

rings

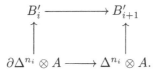

In particular, if $A \longrightarrow B$ is a morphism of commutative rings, being homotopically finitely presented is a much stronger condition than being finitely presented in the usual sense. For instance, if $B$ is a commutative $k$-algebra of finite type (for $k$ a field) which admits a singularity which is not a local complete intersection, then $B$ is not homotopically finitely presented over $k$. Intuitively, $B$ is homotopically finitely presented over $A$ when it admits (up to a retract) a finite resolution by free $A$-algebras of finite type.

• For a given $A \in sComm$, there exists a functor

$$\pi_0 \colon \operatorname{Ho}(A/sComm) \longrightarrow \pi_0(A)/Comm$$

from the homotopy category of commutative simplicial $A$-algebras to the category of commutative $\pi_0(A)$-algebras. This functor induces an equivalence on the full subcategories of étale morphisms and Zariski open immersions. The corresponding fact for smooth and flat morphisms is not true.

**Exercise 4.3.2.** Let $A \longrightarrow B$ be a morphism in $sComm$. Show that $A \longrightarrow B$ is formally étale if and only if $B \otimes_A^{\mathbb{L}} B \longrightarrow B$ is formally smooth.

**Exercise 4.3.3.**   1. Let $A \longrightarrow B$ be a morphsm in $sComm$ which is formally étale and such that $\pi_0(A) \longrightarrow \pi_0(B)$ is an isomorphism. Show that $f$ is an isomorphism in $\operatorname{Ho}(sComm)$.

   2. Deduce from part 1 that, if $A \longrightarrow B$ is formally étale, then the natural morphism

$$B \longrightarrow B \otimes_{B \otimes_A^{\mathbb{L}} B}^{\mathbb{L}} B$$

is an isomorphism in $\operatorname{Ho}(sComm)$.

   3. Let $A \longrightarrow B$ be a morphism in $\operatorname{Ho}(sComm)$ which is formally étale. Show that, for any $C \in A/sComm$, the mapping space $\operatorname{Map}_{A/sComm}(B, C)$ is homotopically discrete (i.e., equivalent to a set).

**Exercise 4.3.4.** Let $f \colon A \longrightarrow B$ be a morphism of non-simplicial commutative rings. Show that $f$ is homotopically of finite presentation in $sComm$ if and only if it is finitely presented as a morphism of rings and $\mathbb{L}_{B/A}$ is a perfect simplicial $B$-module (i.e., corresponds to a perfect complex of $B$-modules by the Dold–Kan correspondence).

# Lecture 5

# Derived stacks and derived algebraic stacks

We arrive at the notions of derived stacks and derived algebraic stacks. We first present the homotopy theory of derived stacks, which is very similar to the homotopy theory of simplicial presheaves presented in §2, but where the category *Comm* must be replaced by the more complicated model category *sComm*. The new feature here is to take into account correctly the model category structure of *sComm*, which makes the definitions a bit more technical.

## 5.1  Derived stacks

We set $dAff = sComm^{\mathrm{op}}$, which is by definition the category of derived affine schemes. It is endowed with the dual model category structure of *sComm*. An object in *dAff* corresponding to $A \in sComm$ will be formally denoted by $\operatorname{Spec} A$. This "Spec" has only a formal meaning, and we will define another, more interesting, Spec functor that will be denoted by $\mathbb{R}\mathrm{Spec}$.

We consider $sPr(dAff)$, the category of simplicial presheaves on *dAff*. We will define three different model category structures on $sPr(dAff)$, each one being a left Bousfield localization of the previous one. In order to avoid confusion, we will use different notations for these three model categories (contrary to what we have done in Lectures 2 and 3), even though the underlying categories are identical. They will be denoted by $sPr(dAff)$, $dAff^{\wedge}$ and $dAff^{\sim}$.

The first model structure is the projective levelwise model category structure on $sPr(dAff)$, for which equivalences and fibrations are defined levelwise. We do not give any specific name to this model category. We consider the Yoneda embedding

$$
\begin{array}{rccc}
h: & dAff & \longrightarrow & sPr(dAff) \\
& X & \longmapsto & h_X = \operatorname{Hom}(-, X).
\end{array}
$$

I. Moerdijk and B. Toën, *Simplicial Methods for Operads and Algebraic Geometry*, Advanced Courses in Mathematics - CRM Barcelona, DOI 10.1007/978-3-0348-0052-5_13, © Springer Basel AG 2010

Here $h_X$ is a presheaf of sets and is considered as a simplicial presheaf constant in the simplicial direction. For any equivalence $X \longrightarrow Y$ in $dAff$ we deduce a morphism $h_X \longrightarrow h_Y$ in $sPr(dAff)$. By definition, the model category $dAff^\wedge$ is the left Bousfield localization of the model category $sPr(dAff)$ with respect to the set of all morphisms $h_X \longrightarrow h_Y$ obtained from equivalences $X \longrightarrow Y$ in $dAff$. The model category $dAff^\wedge$ is called the *model category of prestacks* over $dAff$.

The fibrant objects in $dAff^\wedge$ are the simplicial presheaves $F\colon dAff^{op} \longrightarrow sSet$ satisfying the following two conditions:

1. For any $X \in dAff$, the simplicial set $F(X)$ is fibrant.

2. For any equivalence $X \longrightarrow Y$ in $dAff$, the induced morphism $F(Y) \longrightarrow F(X)$ is an equivalence of simplicial sets.

The first condition above is technical, but the second one is not. This second condition is called the *prestack* condition. This is the essential new feature of derived stack theory compared with stack theory, in which this condition did not appear (simply because there is no notion of equivalence in *Comm* except the trivial one, i.e., the notion of isomorphism). The standard results about left Bousfield localizations imply that $\mathrm{Ho}(dAff^\wedge)$ is naturally equivalent to the full subcategory of $\mathrm{Ho}(sPr(dAff))$ consisting of all simplicial presheaves satisfying condition 2 above. We will implicitly identify these two categories. Moreover, the inclusion functor

$$\mathrm{Ho}(dAff^\wedge) \longrightarrow \mathrm{Ho}(sPr(dAff))$$

has a left adjoint which simply consists in sending a simplicial presheaf $F$ to its fibrant model.

**Exercise 5.1.1.** The natural projection $dAff \longrightarrow \mathrm{Ho}(dAff)$ induces a functor

$$\mathrm{Ho}(sPr(\mathrm{Ho}(dAff))) \longrightarrow \mathrm{Ho}(dAff^\wedge).$$

Show that this functor is not an equivalence of categories.

We come back to the Yoneda functor

$$h\colon dAff \longrightarrow sPr(dAff) = dAff^\wedge.$$

We compose it with the natural functor $dAff^\wedge \longrightarrow \mathrm{Ho}(dAff^\wedge)$ and obtain a functor

$$h\colon dAff \longrightarrow \mathrm{Ho}(dAff^\wedge).$$

By construction, $h$ sends equivalences in $dAff$ to isomorphisms in $\mathrm{Ho}(dAff^\wedge)$. Therefore it induces a well-defined functor

$$\mathrm{Ho}(h)\colon \mathrm{Ho}(dAff) \longrightarrow \mathrm{Ho}(dAff^\wedge).$$

A general result, called the Yoneda lemma for model categories (see [HAGI]), states two properties concerning $\mathrm{Ho}(h)$:

1. The functor $\mathrm{Ho}(h)$ is fully faithful. This is the model category version of the Yoneda lemma for categories.

2. For $X \in dAff$, the object $\mathrm{Ho}(h)(X) \in \mathrm{Ho}(sPr(dAff))$ can be described as follows. We take a fibrant model $R(X)$ for $X$ in $dAff$ (i.e., if $X = \mathrm{Spec}\, A$, then $R(X) = \mathrm{Spec}\, Q(A)$ for $Q(A)$ a cofibrant model for $A$ in $sComm$). We consider the simplicial presheaf $\underline{h}_{R(X)} \colon Y \longmapsto \underline{\mathrm{Hom}}(Y, R(X))$, where $\underline{\mathrm{Hom}}$ denotes the simplicial Hom sets of $dAff$. When $X = \mathrm{Spec}\, A$, this simplicial presheaf is also $\mathrm{Spec}\, B \longmapsto \underline{\mathrm{Hom}}(Q(A), B)$. Then the simplicial presheaf $\mathrm{Ho}(h)(X)$ is equivalent to $\underline{h}_{R(X)}$. When $X = \mathrm{Spec}\, A$, we will also use the following notation:

$$\mathbb{R}\underline{\mathrm{Spec}}\, A = \mathrm{Ho}(h)(X) \simeq \underline{h}_{R(X)} \simeq \underline{\mathrm{Hom}}(Q(A), -).$$

In an equivalent way, the Yoneda lemma in this setting states that the functor $A \longmapsto \underline{\mathrm{Hom}}(Q(A), -)$ induces a fully faithful functor

$$\mathrm{Ho}(Comm)^{\mathrm{op}} \longrightarrow \mathrm{Ho}(dAff^{\wedge}) \subset \mathrm{Ho}(sPr(dAff)).$$

We now introduce the notion of local equivalences for morphisms in $dAff^{\wedge}$ which will be our equivalences for the final model structure. For this, we endow the category $\mathrm{Ho}(dAff)$ with a Grothendieck topology as follows. We say that a family of morphisms $\{A \longrightarrow A_i\}_i$ is an *étale covering* if each of the morphisms $A \longrightarrow A_i$ is étale in the sense of Definition 4.3.1 and if the family of functors

$$\{- \otimes_A^{\mathbb{L}} A_i \colon \mathrm{Ho}(sA\text{-}Mod) \longrightarrow \mathrm{Ho}(sA_i\text{-}Mod)\}_i$$

is conservative. By definition, the étale topology on $\mathrm{Ho}(dAff)$ is the topology for which covering sieves are generated by étale covering families. In the same way, for any fixed $X \in dAff$ we define an étale topology on $\mathrm{Ho}(dAff/X)$.

The étale topology on $\mathrm{Ho}(dAff)$ can be used in order to define homotopy sheaves for objects $F \in \mathrm{Ho}(dAff^{\wedge})$. We start by defining homotopy presheaves as follows. Let $F \colon dAff^{\mathrm{op}} \longrightarrow sSet$ be an object in $\mathrm{Ho}(dAff^{\wedge})$, so in particular we assume that $F$ sends equivalences in $dAff$ to equivalences in $sSet$. We consider the presheaf of sets $X \longmapsto \pi_0(F(X))$. This presheaf sends equivalences in $dAff$ to isomorphisms in $Set$ and thus factorizes as a functor $\pi_0^{\mathrm{pr}}(F) \colon \mathrm{Ho}(dAff)^{\mathrm{op}} \longrightarrow Set$. Similarly, for $X \in dAff$ and $s \in F(X)$, we define a presheaf of groups on $dAff/X$ which sends $f \colon Y \longrightarrow X$ to $\pi_i(F(Y), f^*(s))$. Again this presheaf sends equivalences to isomorphisms and thus induces a functor $\pi_i^{\mathrm{pr}}(F, s) \colon \mathrm{Ho}(dAff/X)^{\mathrm{op}} \longrightarrow Set$. With these notations, the associated sheaves (for the étale topology defined above) to $\pi_0^{\mathrm{pr}}(F)$ and $\pi_i^{\mathrm{pr}}(F, s)$ are denoted by $\pi_0(F)$ and $\pi_i(F, s)$ and are called the *homotopy sheaves* of $F$. These are defined for $F \colon dAff^{\mathrm{op}} \longrightarrow sSet$ sending equivalences to equivalences. Now, for a general simplicial presheaf, we set

$$\pi_0(F) = \pi_0(F^{\wedge}), \qquad \pi_i(F, s) = \pi_i(F^{\wedge}, s),$$

where $F^{\wedge}$ is a fibrant model for $F$ in $dAff^{\wedge}$.

**Definition 5.1.2.** Let $f\colon F \longrightarrow F'$ be a morphism of simplicial presheaves in $dAff$.

1. The morphism $f$ is a *local equivalence* if it satisfies the following two conditions:

   (a) The induced morphism $\pi_0(F) \longrightarrow \pi_0(F')$ is an isomorphism of sheaves in $\mathrm{Ho}(dAff)$.

   (b) For any $X \in dAff$, any $s \in F(X)_0$ and any $i > 0$, the induced morphism $\pi_i(F, s) \longrightarrow \pi_i(F', f(s))$ is an isomorphism of sheaves in $dAff/X$.

2. The morphism $f$ is a *local cofibration* if it is a cofibration in $dAff^\wedge$ (or equivalently in $sPr(dAff)$).

3. The morphism $f$ is a *local fibration* if it has the left lifting property with respect to every local cofibration which is also a local equivalence.

For simplicity, we will use the expressions *equivalence*, *fibration* and *cofibration* in order to refer to *local equivalence*, *local fibration* and *local cofibration*.

It can be proved (see [HAGI]) that these notions of equivalence, fibration and cofibration define a model category structure on $sPr(dAff)$. This model category will be denoted by $dAff^\sim$. As for the case of simplicial presheaves, it is possible to characterize fibrant objects in $dAff^\sim$ as functors $F\colon dAff^{\mathrm{op}} \longrightarrow sSet$ satisfying the following three conditions (we do not make precise the definition of étale hypercovering in this context —it is very similar to the one we gave for simplicial presheaves in §2.1):

1. For any $X \in dAff$, the simplicial set $F(X)$ is fibrant.

2. For any equivalence $X \longrightarrow Y$ in $dAff$, the induced morphism $F(Y) \longrightarrow F(X)$ is an equivalence of simplicial sets.

3. For any $X \in dAff$ and any étale hypercovering $H \longrightarrow X$, the natural morphism
$$F(X) \longrightarrow \mathrm{Holim}_{[n]\in\Delta}\, F(H_n)$$
is an equivalence of simplicial sets.

**Definition 5.1.3.**   1. An object $F \in sPr(dAff)$ is called a *derived stack* if it satisfies the conditions 2 and 3 above.

2. The homotopy category $\mathrm{Ho}(dAff^\sim)$ will be called the *homotopy category of derived stacks*. Most often objects in $\mathrm{Ho}(dAff^\sim)$ will simply be called *derived stacks*. The expressions *morphism of derived stacks* and *isomorphism of derived stacks* will refer to morphisms and isomorphisms in $\mathrm{Ho}(dAff^\sim)$. The set of morphisms of derived stacks from $F$ to $F'$ will be denoted by $[F, F']$.

It can be shown that $dAff^\sim$ is not Quillen equivalent to a model category of the form $sPr(C)$ for any Grothendieck site $C$. The data of $dAff$, together with the étale topology on Ho($dAff$), is therefore a new kind of object, that cannot be recovered from Grothendieck's theory of sites and topoi. The object ($dAff, et$) is called a *model site*, and $dAff^{\sim, et}$ a *model topos*, as they are homotopy-theoretical versions of sites and topoi. We refer to [HAGI] for more about these notions.

To finish this first section, we mention how stacks and derived stacks are compared. For this, we consider the functor $Comm \longrightarrow sComm$ which consists in considering a commutative ring as a constant simplicial commutative ring. This induces a functor $i \colon Aff \longrightarrow dAff$. Pulling back along this functor induces a functor

$$i^* \colon dAff^{\sim} \longrightarrow sPr(Aff).$$

This functor is seen to be right Quillen and its left adjoint is denoted by

$$i_! \colon sPr(Aff) \longrightarrow dAff^{\sim}.$$

The derived Quillen adjunction is denoted by

$$j \colon \mathrm{Ho}(sPr(Aff)) \longrightarrow \mathrm{Ho}(dAff^{\sim}), \qquad \mathrm{Ho}(sPr(Aff)) \longleftarrow \mathrm{Ho}(dAff^{\sim}) \colon h^0.$$

The functor $j$ is fully faithful, as follows from the fact that the functor $Comm \longrightarrow \mathrm{Ho}(sComm)$ is fully faithful and compatible with the étale topologies on both sides. Therefore, any stack can be considered as a derived stack. The functor $h^0$ is called the *classical part functor*, and remembers only the part related to non-simplicial commutative rings of a given derived stack. Using the functor $j$ we will see any stack as a derived stack.

**Definition 5.1.4.** Given a stack $F \in \mathrm{Ho}(sPr(Aff))$, a *derived extension* of $F$ is the data of a derived stack $\widetilde{F} \in \mathrm{Ho}(dAff^{\sim})$ together with an isomorphism of stacks $F \simeq h^0(\widetilde{F})$.

The existence of the full embedding $j$ implies that any stack admits a derived extension $j(F)$, but this extension is somehow the trivial one. The striking fact about derived algebraic geometry is that most (if not all) of the moduli problems admit natural derived extensions, and these are not trivial in general. We will see many such examples in the next lecture.

## 5.2 Algebraic derived $n$-stacks

We now mimic the definitions of schemes and algebraic stacks given in §3 for our new context of derived stacks.

We start by considering the Yoneda embedding

$$\mathrm{Ho}(dAff) \longrightarrow \mathrm{Ho}(dAff^{\wedge}).$$

The faithfully flat descent stays true in the derived setting, and this embedding induces a fully faithful functor

$$\mathrm{Ho}(dAff) \longrightarrow \mathrm{Ho}(dAff^{\sim}).$$

Equivalently, this means that, for any $A \in sComm$, the prestack $\mathbb{R}\mathrm{Spec}\, A$, sending $B$ to $\underline{\mathrm{Hom}}(Q(A), B)$, satisfies the descent condition for étale hypercoverings (i.e., it is a derived stack). Objects in the essential image of this functor will be called *derived affine schemes*, and the full subcategory of $\mathrm{Ho}(dAff^\sim)$ consisting of derived affine schemes will be implicitly identified with $\mathrm{Ho}(dAff)$.

One of the major differences between stacks and derived stacks is that derived affine schemes are not 0-truncated. The definition of Zariski open immersion given in Definition 3.1.1 has therefore to be slightly modified.

**Definition 5.2.1.**    1. A morphism of derived stacks $F \longrightarrow F'$ is a *monomorphism* if the induced morphism $F \longrightarrow F \times^h_{F'} F$ is an equivalence.

2. A morphism of derived stacks $F \longrightarrow F'$ is an *epimorphism* if the induced morphism $\pi_0(F) \longrightarrow \pi_0(F')$ is an epimorphism of sheaves.

3. Let $X = \mathbb{R}\mathrm{Spec}\, A$ be a derived affine scheme, $F$ a derived stack and $i: F \longrightarrow \mathbb{R}\mathrm{Spec}\, A$ a morphism. We say that $i$ is a *Zariski open immersion* (or simply an *open immersion*) if it satisfies the following two conditions:

   (a) The morphism $i$ is a monomorphism.

   (b) There exists a family of Zariski open immersions $\{A \longrightarrow A_i\}_i$ such that the morphisms $\mathbb{R}\mathrm{Spec}\, A_i \longrightarrow \mathbb{R}\mathrm{Spec}\, A$ all factor through $F$ in a way that the resulting morphism

   $$\coprod_i \mathbb{R}\mathrm{Spec}\, A_i \longrightarrow \mathbb{R}\mathrm{Spec}\, A$$

   is an epimorphism.

4. A morphism of derived stacks $F \longrightarrow F'$ is a *Zariski open immersion* (or simply an *open immersion*) if, for any derived affine scheme $X$ and any morphism $X \longrightarrow F'$, the induced morphism

   $$F \times^h_{F'} X \longrightarrow X$$

   is a Zariski open immersion in the above sense.

5. A derived stack $F$ is a *derived scheme* if there exist a family of derived affine schemes $\{\mathbb{R}\mathrm{Spec}\, A_i\}_i$ and Zariski open immersions $\mathbb{R}\mathrm{Spec}\, A_i \longrightarrow F$ such that the induced morphism of sheaves

   $$\coprod_i \mathbb{R}\mathrm{Spec}\, A_i \longrightarrow F$$

   is an epimorphism. Such a family of morphisms $\{\mathbb{R}\mathrm{Spec}\, A_i \longrightarrow F\}$ will be called a *Zariski atlas* for $F$.

We say that a morphism of derived schemes $X \longrightarrow Y$ is smooth (resp. flat or étale) if there exist Zariski atlases $\{\mathbb{R}\mathrm{Spec}\, A_i \longrightarrow X\}$ and $\{\mathbb{R}\mathrm{Spec}\, A_j \longrightarrow Y\}$ together with commutative squares in $\mathrm{Ho}(dAff^{\sim})$

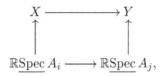

with $\mathbb{R}\mathrm{Spec}\, A_i \longrightarrow \mathbb{R}\mathrm{Spec}\, A_j$ a smooth (resp. flat or étale) morphism —here $j$ depends on $i$. Smooth morphisms of derived schemes are stable under composition and homotopy base change.

The following is the main definition of this series of lectures.

**Definition 5.2.2.**    1. A derived stack $F$ is 0-*algebraic* if it is a derived scheme.

2. A morphism of derived stacks $F \longrightarrow F'$ is 0-*algebraic* (or 0-*representable*) if, for any derived scheme $X$ and any morphism $X \longrightarrow F'$, the derived stack $F \times_{F'}^h X$ is 0-algebraic (i.e., a derived scheme).

3. A 0-algebraic morphism of derived stacks $F \longrightarrow F'$ is smooth if, for any derived scheme $X$ and any morphism $X \longrightarrow F'$, the morphism of derived schemes $F \times_{F'}^h X \longrightarrow X$ is smooth.

4. We now let $n > 0$ and assume that the notions of $(n-1)$-algebraic derived stack, $(n-1)$-algebraic morphism and smooth $(n-1)$-algebraic morphism have been defined.

    (a) A derived stack $F$ is $n$-*algebraic* if there exists a derived scheme $X$ together with a smooth $(n-1)$-algebraic morphism $X \longrightarrow F$ which is an epimorphism. Such a morphism $X \longrightarrow F$ is called a *smooth $n$-atlas* for $F$.

    (b) A morphism of derived stacks $F \longrightarrow F'$ is $n$-*algebraic* (or $n$-*representable*) if, for any derived scheme $X$ and any morphism $X \longrightarrow F'$, the derived stack $F \times_{F'} X$ is $n$-algebraic.

    (c) An $n$-algebraic morphism of derived stacks $F \longrightarrow F'$ is *smooth* (resp. *flat* or *étale*) if, for any derived scheme $X$ and any morphism $X \longrightarrow F'$, there exists a smooth $n$-atlas $Y \longrightarrow F \times_{F'}^h X$ such that each morphism $Y \longrightarrow X$ is a smooth (resp. flat or étale) morphism of derived schemes.

5. A *derived algebraic stack* is a derived stack which is $n$-algebraic for some $n$.

6. A morphism of derived stacks $F \longrightarrow F'$ is *algebraic* (or *representable*) if it is $n$-algebraic for some $n$.

7. A morphism of derived stacks $F \longrightarrow F'$ is *smooth* (resp. *flat* or *étale*) if it is $n$-algebraic and smooth (resp. flat or étale) for some $n$.

We finish this part with some basic properties of derived algebraic stacks, and in particular with a comparison between the notions of algebraic stacks and derived algebraic stacks.

- Derived algebraic stacks are stable under finite homotopy limits (i.e., homotopy pullbacks).

- Derived algebraic stacks are stable under disjoint union.

- Algebraic morphisms of derived stacks are stable under composition and homotopy base change.

- Derived algebraic stacks are stable under smooth quotients. To be more precise, if $F \longrightarrow F'$ is a smooth epimorphism of derived stacks, then $F'$ is algebraic if and only if $F$ is so.

- A (non-derived) stack $F$ is algebraic if and only if the derived stack $j(F)$ is algebraic.

- If $F$ is an algebraic derived stack, then the stack $h^0(F)$ is an algebraic stack. When $h^0(F)$ is an algebraic $n$-stack, we say that $F$ is a derived algebraic $n$-stack (although it is not $n$-truncated as a simplicial presheaf on $dAff$).

- A derived algebraic space is a derived algebraic stack $F$ such that $h^0(F)$ is an algebraic space. In other words, a derived algebraic space is a derived algebraic 0-stack.

- If $F$ is an algebraic derived $n$-stack and $A$ is an $m$-truncated commutative simplicial ring, then $F(A)$ is an $(n+m)$-truncated simplicial set.

- If $f \colon F \longrightarrow F'$ is a flat morphism of derived algebraic stacks, and if $F'$ is an algebraic stack (i.e., of the form $j(F'')$ for an algebraic stack $F''$), then $F$ is itself an algebraic stack.

We see that the formal properties of derived algebraic stacks are the same as the formal properties of non-derived algebraic stacks. However, we would like to make the important comment here that the inclusion functor $j \colon \mathrm{Ho}(sPr(Aff)) \longrightarrow \mathrm{Ho}(dAff^\sim)$ from stacks to derived stacks does not commute with homotopy pullbacks. In other words, if $F \longleftarrow H \longrightarrow G$ is a diagram of stacks, then the natural morphism

$$j(F \times^h_H G) \longrightarrow j(F) \times^h_{j(H)} j(G)$$

is not an isomorphism in general. As this morphism induces an isomorphism on $h^0$, this is an example of a non-trivial derived extension of a stack as a derived stack. Each time a stack is presented as a certain finite homotopy limit of other stacks, it has a natural (and in general non-trivial) derived extension by considering the same homotopy limit in the bigger category of derived stacks.

## 5.3 Cotangent complexes

To finish this lecture, we now explain the notion of cotangent complexes of a derived stack at a given point. We let $F$ be an algebraic derived stack and $X = \operatorname{Spec} A$ be a (non-derived) affine scheme. We fix a point (i.e., a morphism of stacks)

$$x \colon X \longrightarrow F.$$

We let $D^{\leq 0}(A)$ be the non-positive derived category of cochain complexes of $A$-modules. By the Dold–Kan correspondence, we will also identify $D^{\leq 0}(A)$ with the homotopy category $\operatorname{Ho}(sA\text{-}Mod)$ of simplicial $A$-modules. We define a functor

$$\mathbb{D}er_x(F, -) \colon D^{\leq 0}(A) \longrightarrow \operatorname{Ho}(sSet)$$

in the following way. For $M \in D^{\leq 0}(A)$, we form $A \oplus M$, which is now a commutative simplicial ring (here we consider $M$ as a simplicial $A$-module), and we set $X[M] = \mathbb{R}\underline{\operatorname{Spec}}(A \oplus M)$. The natural projection $A \oplus M \longrightarrow A$ induces a morphism of derived schemes $X \longrightarrow X[M]$. By definition, the simplicial set $\mathbb{D}er_x(F, M)$ is the homotopy fiber of the natural morphism

$$F(X[M]) \longrightarrow F(X)$$

taken at the point $x$ (here we use the Yoneda lemma, stating that $\pi_0(F(X)) \simeq [X, F]$). The simplicial set $\mathbb{D}er_x(F, M)$ is called the simplicial set of derivations of $F$ at the point $x$ with coefficients in $M$. This is functorial in $M$ and thus defines a functor

$$\mathbb{D}er_x(F, -) \colon D^{\leq 0}(A) \longrightarrow \operatorname{Ho}(sSet).$$

It can be proved that this functor is corepresentable by a complex of $A$-modules. More precisely, there exists a complex of $A$-modules (a priori not concentrated in non-positive degrees any more) $\mathbb{L}_{F,x}$, called the cotangent complex of $F$ at $x$, and such that there exist natural isomorphisms in $\operatorname{Ho}(sSet)$,

$$\mathbb{D}er_x(F, M) \simeq \operatorname{Map}(\mathbb{L}_{F,x}, M),$$

where Map are the mapping spaces of the model category of (unbounded) complexes of $A$-modules. When the derived stack $F$ is affine, this is a reformulation of the existence of a cotangent complex as recalled in §4. In general, one reduces the statement to the affine case by a long and tedious induction (on $n$, proving the result for algebraic derived $n$-stacks). Finally, with a bit of care, we can show that $\mathbb{L}_{F,x}$ is unique and functorial —although this requires us to state a refined universal property; see [HAGII].

**Definition 5.3.1.** With notation as above, the complex $\mathbb{L}_{F,x}$ is called the *cotangent complex* of $F$ at the point $x$. Its dual $\mathbb{T}_{F,x} = \mathbb{R}\underline{\operatorname{Hom}}(\mathbb{L}_{F,x}, A)$ is called the *tangent complex* of $F$ at $x$. The cohomology groups

$$T_{F,x}^i = H^i(\mathbb{T}_{F,x})$$

are called the *higher tangent spaces* of $F$ at $x$.

For a derived algebraic $n$-stack $F$ and a point $x\colon X = \operatorname{Spec} A \longrightarrow F$, the cotangent complex $\mathbb{L}_{F,x}$ belongs to the derived category $D^{\leq n}(A)$ of complexes concentrated in degrees $]-\infty, n]$. The part of $\mathbb{L}_{F,x}$ concentrated in negative degrees is the one related to the *derived part* of $F$ (i.e., the one making the difference between commutative rings and commutative simplicial rings), and the non-negative part is related to the *stacky part* of $F$ (i.e., the part related to the higher homotopy sheaves of $h^0(F)$). For instance, when $F$ is a derived scheme, its stacky part is trivial and thus $\mathbb{L}_{F,x}$ belongs to $D^{\leq 0}(A)$. On the other hand, when $F$ is a smooth algebraic $n$-stack (say over $\operatorname{Spec} \mathbb{Z}$ to simplify), then $\mathbb{L}_{F,x}$ is concentrated in degrees $[0, n]$ —even more is true: it is of Tor-amplitude concentrated in degrees $[0, n]$.

In the next lecture we will give several examples of derived stacks which will show that the tangent complexes contain interesting cohomological information. Also, tangent complexes are very useful to provide smoothness and étaleness criteria, which are in general easy to check in practice. For this reason, proving smoothness is in general much easier in the context of derived algebraic geometry than in the usual context of algebraic geometry. Here is, for instance, a smoothness criterion (see [HAGII] for details).

Let $f\colon F \longrightarrow F'$ be a morphism of algebraic derived stacks. For any affine scheme $X = \operatorname{Spec} A$ and any point $x\colon X \longrightarrow F$, we consider the homotopy fiber of the natural morphism

$$\mathbb{L}_{F,x} \longrightarrow \mathbb{L}_{F',x},$$

which is called the relative cotangent complex of $f$ at $x$ and is denoted by $\mathbb{L}_{f,x}$. We assume that $f$ is locally homotopically of finite presentation (i.e., $F$ and $F'$ admit atlases compatible with $f$ such that the induced morphism on the atlases is homotopically of finite presentation). Then $f$ is smooth if and only if, for any affine scheme $X = \operatorname{Spec} A$ and any $x\colon X \longrightarrow F$, the complex $\mathbb{L}_{f,x}$ is of non-negative Tor amplitude (i.e., for all $M \in D^{\leq -1}(A)$ we have $[\mathbb{L}_{f,x}, M] = 0$).

**Exercise 5.3.2.** Show that a morphism of derived algebraic stacks $f\colon F \longrightarrow F'$ is an isomorphism in $\operatorname{Ho}(dAff^{\sim})$ if and only if it satisfies the following three conditions:

1. The morphism is locally homotopically finitely presented.

2. For all fields $K$, the induced morphism

$$F(K) \longrightarrow G(K)$$

   is an equivalence.

3. For all fields $K$ and all morphisms $x\colon \operatorname{Spec} K \longrightarrow F$, we have $\mathbb{L}_{f,x} \simeq 0$.

**Exercise 5.3.3.** By definition, a derived scheme is a derived algebraic stack $F$ such that $h^0(F)$ is a scheme.

1. Show the existence of an equivalence between the small Zariski site of $F$ and the one of $h^0(F)$.

2. For $U = \mathbb{R}\mathrm{Spec}\, A \longrightarrow F$ an open Zariski immersion, we let $\pi_i^{\mathrm{virt}}(U) = \pi_i(A)$, which is a $\pi_0(A)$-module. Show that $U \longmapsto \pi_i^{\mathrm{virt}}(U)$ defines quasi-coherent sheaves on the scheme $h^0(F)$.

3. Suppose that $h^0(F)$ is now a locally noetherian scheme. Assume that, for any Zariski open immersion $\mathbb{R}\mathrm{Spec}\, A \longrightarrow F$, the homotopy groups $\pi_i(A)$ are of finite type and vanish for $i$ big enough. Show that the class

$$[F]^{\mathrm{virt}} = \sum_i (-1)^i [\pi_i^{\mathrm{virt}}] \in G_0(h^0(F))$$

is a well-defined class in the Grothendieck group of coherent sheaves. It is called the *virtual K-theory fundamental class* of $F$.

# Lecture 6

# Examples of derived algebraic stacks

In this last lecture, we present examples of derived algebraic stacks.

## 6.1 The derived moduli space of local systems

We come back to the example that we presented in the first lecture, namely the moduli problem of linear representations of a discrete group. We will now reconsider it from the point of view of derived algebraic geometry. We will try to treat this example in some detail, as we think it is a rather simple, but interesting, example of a derived algebraic stack.

A linear representation of a group $G$ can also be interpreted as a local system on the space $BG$. We will therefore study the moduli problem from this topological point of view. We fix a finite CW-complex $X$ and we are going to define a derived stack $\mathbb{R}\mathrm{Loc}(X)$ classifying local systems on $X$. We will see that this stack is an algebraic derived 1-stack and we will describe its higher tangent spaces in terms of cohomology groups of $X$. When $X = BG$ for a discrete group $G$, the derived algebraic stack $\mathbb{R}\mathrm{Loc}(X)$ is the *correct moduli space* of linear representations of $G$.

We start by considering the non-derived algebraic 1-stack **Vect** classifying projective modules of finite type. By definition, **Vect** sends a commutative ring $A$ to the nerve of the groupoid of projective $A$-modules of finite type. The stack **Vect** is a 1-stack. It is easy to see that **Vect** is an algebraic 1-stack. Indeed, we have a decomposition

$$\mathbf{Vect} \simeq \coprod_n \mathbf{Vect}_n,$$

where $\mathbf{Vect}_n \subset \mathbf{Vect}$ is the substack of projective modules of rank $n$ (recall that a projective $A$-module of finite type $M$ is of rank $n$ if, for any field $K$ and any

I. Moerdijk and B. Toën, *Simplicial Methods for Operads and Algebraic Geometry*, Advanced Courses in Mathematics - CRM Barcelona, DOI 10.1007/978-3-0348-0052-5_14, © Springer Basel AG 2010

morphism $A \longrightarrow K$, the $K$-vector space $M \otimes_A K$ is of dimension $n$). It is therefore enough to prove that $\mathbf{Vect}_n$ is an algebraic 1-stack. This last statement will itself follow from the identification

$$\mathbf{Vect}_n \simeq [*/Gl_n] = BGl_n,$$

where $Gl_n$ is the affine group scheme sending $A$ to $Gl_n(A)$. In order to prove that $\mathbf{Vect}_n \simeq BGl_n$, we construct a morphism of simplicial presheaves

$$BGl_n \longrightarrow \mathbf{Vect}_n$$

by sending the base point of $BGl_n$ to the trivial projective module of rank $n$. For a given commutative ring $A$, the morphism

$$BGl_n(A) \longrightarrow \mathbf{Vect}_n(A)$$

sends the base point to $A^n$ and identifies $Gl_n(A)$ with the automorphism group of $A^n$. The claim is that the morphism $BGl_n \longrightarrow \mathbf{Vect}_n$ is a local equivalence of simplicial presheaves. As, by construction, this morphism induces isomorphisms on all higher homotopy sheaves, it only remains to show that it induces an isomorphism on the sheaves $\pi_0$. But this in turn follows from the fact that $\pi_0(\mathbf{Vect}_n) \simeq *$, because any projective $A$-module of finite type is locally free for the Zariski topology on Spec $A$.

The algebraic stack $\mathbf{Vect}$ is now considered as an algebraic derived stack using the inclusion functor $j\colon \mathrm{Ho}(sPr(Aff)) \longrightarrow \mathrm{Ho}(dAff^\sim)$. We consider a fibrant model $F \in dAff^\sim$ for $j(\mathbf{Vect})$, and we define a new simplicial presheaf

$$\mathbb{R}\mathrm{Loc}(X)\colon dAff^{\mathrm{op}} \longrightarrow sSet$$

which sends $A \in sComm$ to $\mathrm{Map}(X, |F(A)|)$, the simplicial set of continuous maps from $X$ to $|F(A)|$.

**Definition 6.1.1.** The derived stack $\mathbb{R}\mathrm{Loc}(X)$ defined above is called the *derived moduli stack of local systems* on $X$.

We will now describe some basic properties of the derived stack $\mathbb{R}\mathrm{Loc}(X)$. We start by a description of its classical part $h^0(\mathbb{R}\mathrm{Loc}(X))$, which will show that it does classify local systems on $X$. We will then show that $\mathbb{R}\mathrm{Loc}(X)$ is an algebraic derived stack locally of finite presentation over Spec $\mathbb{Z}$, and that it can be written as

$$\mathbb{R}\mathrm{Loc}(X) \simeq \coprod_n \mathbb{R}\mathrm{Loc}_n(X)$$

where $\mathbb{R}\mathrm{Loc}_n(X)$ is the part classifiying local systems of rank $n$ and is itself strongly of finite type. Finally, we will compute its tangent spaces in terms of the cohomology of $X$.

For $A \in Comm$, note that $h^0(\mathbb{R}\mathrm{Loc}(X))(A)$ is by definition the simplicial set $\mathrm{Map}(X, |F(A)|)$. Now, $F(A)$ is a fibrant model for $j(\mathbf{Vect})(A) \simeq \mathbf{Vect}(A)$, and so it is equivalent to the nerve of the groupoid of projective $A$-modules of finite rank. The simplicial set $\mathrm{Map}(X, |F(A)|)$ is then naturally equivalent to the nerve of the groupoid of functors $\mathrm{Fun}(\Pi_1(X), F(A))$ from the fundamental groupoid of $X$ to $F(A)$. This last groupoid is in turn equivalent to the groupoid of local systems of projective $A$-modules of finite type on the space $X$. Thus, we see that $h^0(\mathbb{R}\mathrm{Loc}(X))(A)$ is naturally equivalent to the nerve of the groupoid of local systems of projective $A$-modules of finite type on the space $X$. We thus have the following properties:

1. The set $\pi_0(h^0(\mathbb{R}\mathrm{Loc}(X))(A))$ is functorially in bijection with the set of isomorphism classes of local systems of projective $A$-modules of finite type on $X$. In particular, when $A$ is a field this is also the set of local systems of finite-dimensional vector spaces over $X$.

2. For a local system $E \in \pi_0(h^0(\mathbb{R}\mathrm{Loc}(X))(A))$, we have

$$\pi_1(h^0(\mathbb{R}\mathrm{Loc}(X))(A), E) = \mathrm{Aut}(E),$$

   the automorphism group of $E$ as a sheaf of $A$-modules on $X$.

3. For all $i > 1$ and all $E \in \pi_0(h^0(\mathbb{R}\mathrm{Loc}(X))(A))$, we have

$$\pi_i(h^0(\mathbb{R}\mathrm{Loc}(X))(A), E) = 0.$$

Let us explain now why the derived stack $\mathbb{R}\mathrm{Loc}(X)$ is algebraic. We start with the trivial case where $X$ is a contractible space. Then, by definition, we have $\mathbb{R}\mathrm{Loc}(X) \simeq \mathbb{R}\mathrm{Loc}(*) \simeq j(\mathbf{Vect})$. As we already know that $j(\mathbf{Vect})$ is an algebraic stack, this implies that $\mathbb{R}\mathrm{Loc}(X)$ is an algebraic derived stack when $X$ is contractible.

The next step is to prove that $\mathbb{R}\mathrm{Loc}(S^n)$ is algebraic for any $n \geq 0$. This can be seen by induction on $n$. The case $n = 0$ is obvious. Moreover, for any $n > 0$ we have a homotopy pushout of topological spaces

$$\begin{array}{ccc} S^{n-1} & \longrightarrow & D^n \\ \downarrow & & \downarrow \\ D^n & \longrightarrow & S^n, \end{array}$$

where $D^n$ is the $n$-dimensional ball. This implies the existence of a homotopy pullback diagram of derived stacks

$$\begin{array}{ccc} \mathbb{R}\mathrm{Loc}(S^n) & \longrightarrow & \mathbb{R}\mathrm{Loc}(D^n) \\ \downarrow & & \downarrow \\ \mathbb{R}\mathrm{Loc}(D^n) & \longrightarrow & \mathbb{R}\mathrm{Loc}(S^{n-1}). \end{array}$$

By induction on $n$ and by what we have just seen, the derived stacks $\mathbb{R}\mathrm{Loc}(D^n)$ and $\mathbb{R}\mathrm{Loc}(S^{n-1})$ are algebraic. By the stability of algebraic derived stacks by homotopy pullbacks, we deduce that $\mathbb{R}\mathrm{Loc}(S^n)$ is an algebraic derived stack.

We are now ready to show that $\mathbb{R}\mathrm{Loc}(X)$ is algebraic. We write $X_k$ to denote the $k$-th skeleton of $X$. Since $X$ is a finite CW-complex, there is an $n$ such that $X = X_n$. Moreover, for any $k$ there exists a homotopy pushout diagram of topological spaces

$$\begin{array}{ccc} \coprod S^{k-1} & \longrightarrow & \coprod D^k \\ \downarrow & & \downarrow \\ X_{k-1} & \longrightarrow & X_k, \end{array}$$

where the disjoint unions are finite. This implies that we have a homotopy pullback square of derived stacks

$$\begin{array}{ccc} \mathbb{R}\mathrm{Loc}(X_k) & \longrightarrow & \mathbb{R}\mathrm{Loc}(X_{k-1}) \\ \downarrow & & \downarrow \\ \prod^h \mathbb{R}\mathrm{Loc}(D^k) & \longrightarrow & \prod^h \mathbb{R}\mathrm{Loc}(S^{k-1}). \end{array}$$

By the stability of algebraic derived stacks by finite homotopy limits, we deduce that $\mathbb{R}\mathrm{Loc}(X_k)$ is algebraic by induction on $k$ (the case $k = 0$ being clear, as $\mathbb{R}\mathrm{Loc}(X_0)$ is a finite product of $\mathbb{R}\mathrm{Loc}(*)$).

To finish the study of this example, we will compute the higher tangent spaces of the derived stack $\mathbb{R}\mathrm{Loc}(X)$. We let $A$ be a commutative algebra and consider the natural morphism

$$\mathbb{R}\mathrm{Loc}(*)(A \oplus A[i]) \longrightarrow \mathbb{R}\mathrm{Loc}(*)(A).$$

This morphism has a natural section and its homotopy fiber at an $A$-module $E$ is equivalent to $K(\mathrm{End}(E), i+1)$. It is therefore naturally equivalent to

$$[K(\mathrm{End}(-), i+1)/\mathrm{Vect}(A)] \longrightarrow N(\mathrm{Vect}(A)),$$

where $\mathrm{Vect}(A)$ is the groupoid of projective $A$-modules of finite type, $N(\mathrm{Vect}(A))$ is its nerve, and $[K(\mathrm{End}(-), i+1)/\mathrm{Vect}(A)]$ is the homotopy colimit of the simplicial presheaf $\mathrm{Vect}(A) \longrightarrow sSet$ sending $E$ to $K(\mathrm{End}(E), i+1)$ —this is a general fact: for any simplicial presheaf $F\colon I \longrightarrow sSet$ we have a natural morphism $\mathrm{Hocolim}_I F \longrightarrow N(I) \simeq \mathrm{Hocolim}_I(*)$. We consider the geometric realization of this morphism to get a map of topological spaces

$$|[K(\mathrm{End}(-), i+1)/\mathrm{Vect}(A)]| \longrightarrow |N(\mathrm{Vect}(A))|,$$

which is equivalent to the geometric realization of

$$\mathbb{R}\mathrm{Loc}(*)(A \oplus A[i]) \longrightarrow \mathbb{R}\mathrm{Loc}(*)(A).$$

We take the image of this morphism by $\mathrm{Map}(X, -)$ to get

$$\mathbb{R}\mathrm{Loc}(X)(A \oplus A[i]) \simeq \mathrm{Map}(X, \mathbb{R}\mathrm{Loc}(*)(A \oplus A[i])) \longrightarrow$$
$$\mathrm{Map}(X, \mathbb{R}\mathrm{Loc}(*)(A)) \simeq \mathbb{R}\mathrm{Loc}(X)(A).$$

This implies that the morphism

$$\mathbb{R}\mathrm{Loc}(*)(A \oplus A[i]) \longrightarrow \mathbb{R}\mathrm{Loc}(*)(A)$$

is equivalent to the morphism

$$\mathrm{Map}(X, |[K(\mathrm{End}(-), i+1)/\mathrm{Vect}(A)]|) \longrightarrow \mathrm{Map}(X, |N(\mathrm{Vect}(A))|).$$

A morphism $X \longrightarrow |N(\mathrm{Vect}(A))|$ corresponds to a local system $E$ of projective $A$-modules of finite type on $X$. The homotopy fiber of the above morphism at $E$ is then equivalent to the simplicial set of homotopy lifts of $X \longrightarrow |N(\mathrm{Vect}(A))|$ to a morphism $X \longrightarrow |[K(\mathrm{End}(-), i+1)/\mathrm{Vect}(A)]|$. This simplicial set is in turn naturally equivalent to $DK(C^*(X, \mathrm{End}(E))[i+1])$, the simplicial set obtained from the complex $C^*(X, \mathrm{End}(E))[i+1]$ by the Dold–Kan construction. Here $C^*(X, \mathrm{End}(E))$ denotes the complex of cohomology of $X$ with coefficients in the local system $\mathrm{End}(E)$. We therefore have the following formula for the higher tangent complexes:

$$T_E^i \mathbb{R}\mathrm{Loc}(X) \simeq H^0(C^*(X, \mathrm{End}(E))[i+1]) \simeq H^{i+1}(X, \mathrm{End}(E)).$$

More generally, it is possible to prove that there is an isomorphism in $D(A)$

$$T_E \mathbb{R}\mathrm{Loc}(X) \simeq C^*(X, \mathrm{End}(E))[1].$$

## 6.2   The derived moduli of maps

As for non-derived stacks, the homotopy category of derived stacks $\mathrm{Ho}(dA\!f\!f^{\sim})$ is cartesian closed. The corresponding internal Hom will be denoted by $\mathbb{R}\mathcal{H}om$. Note that, even though we use the same notations for the internal Homs of stacks and derived stacks, the inclusion functor

$$j \colon \mathrm{Ho}(sPr(A\!f\!f)) \longrightarrow \mathrm{Ho}(dA\!f\!f^{\sim})$$

does not commute with them. However, we always have

$$h^0(\mathbb{R}\mathcal{H}om(F, F')) \simeq \mathbb{R}\mathcal{H}om(h^0(F), h^0(F'))$$

for all derived stacks $F$ and $F'$. The situation is therefore very similar to the case of homotopy pullbacks.

We have just seen an example of a derived stack constructed as an internal Hom between two stacks. Indeed, if we use again the notations of the last example, we have

$$\mathbb{R}\mathrm{Loc}(X) \simeq \mathbb{R}\mathcal{H}om(K, \mathbf{Vect}),$$

where $K = S_*(X)$ is the singular simplicial set of $X$.

We now consider another example. Let $X$ and $Y$ be two schemes, and assume that $X$ is flat and proper (say over $\operatorname{Spec} k$ for some base ring $k$), and that $Y$ is smooth over $k$. It is possible to prove that the derived stack

$$\mathbb{R}\underline{\mathcal{H}om}_{dAff/\operatorname{Spec} k}(X, Y)$$

is a derived scheme which is homotopically finitely presented over $\operatorname{Spec} k$. We will not sketch the argument here, as it is out of the scope of these lectures, and we refer to [HAGII] for more details. The derived scheme $\mathbb{R}\underline{\mathcal{H}om}(X, Y)$ is called the *derived moduli space of maps* from $X$ to $Y$. Its classical part $h^0(\mathbb{R}\underline{\mathcal{H}om}(X, Y))$ is the usual moduli scheme of maps from $X$ to $Y$, and for such a map we have

$$\mathbb{T}_f \mathbb{R}\underline{\mathcal{H}om}_{dAff/\operatorname{Spec} k}(X, Y) \simeq C^*(X, f^*(\mathbb{T}_Y)),$$

where all these tangent complexes are relative to $\operatorname{Spec} k$.

We mention here that these derived mapping spaces of maps can also be used in order to construct the so-called derived moduli of stable maps to an algebraic variety, by letting $X$ vary in the moduli space of stable curves. We refer to [To1] for more details about this construction, and for some explanations of how Gromov–Witten theory can be extracted from this derived stack of stable maps.

# Bibliography

[An] M. Anel, Moduli of linear and abelian categories, preprint, arXiv:math. AG/0607385.

[SGA1] M. Artin and A. Grothendieck, *Revêtements étales et groupe fondamental*, Lecture Notes in Math., vol. 224, Springer-Verlag, 1971.

[Ar-Ma] M. Artin and B. Mazur, *Étale Homotopy*, Lecture Notes in Math., vol. 100, Springer-Verlag, 1969.

[Bl] B. Blander, Local projective model structure on simplicial presheaves, *K-theory* **24** (2001), no. 3, 283–301.

[Du1] D. Dugger, Universal homotopy theories, *Adv. Math.* **164** (2001), 144–176.

[DHI] D. Dugger, S. Hollander, and D. Isaksen, Hypercovers and simplicial presheaves, *Math. Proc. Cambridge Philos. Soc.* **136** (2004), 9–51.

[Go-Ho] P. Goerss and M. Hopkins, André–Quillen (co)homology for simplicial algebras over simplicial operads, in: *Une Dégustation Topologique [Topological Morsels]: Homotopy Theory in the Swiss Alps* (D. Arlettaz and K. Hess, eds.), Contemp. Math., vol. 265, Amer. Math. Soc., Providence, 2000, pp. 41–85.

[Hi] P. Hirschhorn, *Model Categories and Their Localizations*, Math. Surveys and Monographs, vol. 99, Amer. Math. Soc., Providence, 2003.

[H-S] A. Hirschowitz and C. Simpson, Descente pour les *n*-champs, preprint, arXiv:math.AG/9807049.

[Hol] S. Hollander, A homotopy theory for stacks, *Israel J. Math.* **163** (2008), 93–124.

[Ho] M. Hovey, *Model Categories*, Math. Surveys and Monographs, vol. 63, Amer. Math. Soc., Providence, 1998.

[Ja] R. Jardine, Simplicial presheaves, *J. Pure Appl. Algebra* **47** (1987), 35–87.

[La-Mo] G. Laumon and L. Moret-Bailly, *Champs algébriques*, Ergeb. Math. Grenzgeb., 3rd Series, vol. 39, Springer-Verlag, 2000.

I. Moerdijk and B. Toën, *Simplicial Methods for Operads and Algebraic Geometry*, Advanced Courses in Mathematics - CRM Barcelona, DOI 10.1007/978-3-0348-0052-5, © Springer Basel AG 2010

[Pr]  J. P. Pridham, Presenting higher stacks as simplicial schemes, preprint, arXiv:0905.4044.

[Q]  D. Quillen, On the (co-)homology of commutative rings, in: *Applications of Categorical Algebra* (Proc. Sympos. Pure Math., vol. XVII, New York, 1968), Amer. Math. Soc., Providence, 1970, pp. 65–87.

[Re]  C. Rezk, Fibrations and homotopy colimits of simplicial sheaves, preprint available at www.math.uiuc.edu/~rezk/papers.html.

[To1]  B. Toën, Higher and derived stacks: A global overview, in: *Algebraic Geometry, Seattle 2005* (Proc. Sympos. Pure Math., vol. 80, part 1), Amer. Math. Soc., Providence, 2009, pp. 435–487.

[To2]  B. Toën, Derived Hall algebras, *Duke Math. J.* **135** (2006), no. 3, 587–615.

[To-Va]  B. Toën and M. Vaquié, Moduli of objects in dg-categories, *Ann. Sci. École Norm. Sup.* **40** (2007), no. 3, pp. 387–444.

[HAGI]  B. Toën and G. Vezzosi, Homotopical algebraic geometry I: Topos theory, *Adv. Math.* **193** (2005), 257–372.

[HAGII]  B. Toën and G. Vezzosi, *Homotopical Algebraic Geometry II: Geometric Stacks and Applications*, Mem. Amer. Math. Soc., vol. 193 (2008), no. 902.